APPROACH TO
PHYSIOLOGY

a thinking APPROACH TO PHYSIOLOGY

Ian N. Sabir

MA PhD
Bye-Fellow in Physiology, Downing College, University of Cambridge

Juliet A. Usher-Smith

MA MB BChir PhD
Fellow in Physiology, New Hall, University of Cambridge

 World Scientific

NEW JERSEY • LONDON • SINGAPORE • BEIJING • SHANGHAI • HONG KONG • TAIPEI • CHENNAI

Published by

World Scientific Publishing Co. Pte. Ltd.

5 Toh Tuck Link, Singapore 596224

USA office: 27 Warren Street, Suite 401-402, Hackensack, NJ 07601

UK office: 57 Shelton Street, Covent Garden, London WC2H 9HE

British Library Cataloguing-in-Publication Data
A catalogue record for this book is available from the British Library.

ISBN-13 978-981-270-601-0
ISBN-10 981-270-601-1
ISBN-13 978-981-270-602-7 (pbk)
ISBN-10 981-270-602-X (pbk)

Typeset by Stallion Press
E-mail: enquiries@stallionpress.com

Printed in Singapore.

Foreword

Therefore, since brevity is the soul of wit,
And tediousness the limbs and outward flourishes,
I will be brief.

Hamlet (Act II, scene ii)
W. Shakespeare

There is no shortage of major textbooks on medicine or the biological sciences that comprehensively cover the facts and background of Physiology. This book has a different aim. It is written to smooth the paths of medical and science students on the eve of their ventures into the fascinating world of their chosen subject and profession. For those who have already embarked on this journey, it provides a thoughtful and reflective view of Physiology. It has a friendly tone, is brief, and seeks to introduce and explain, rather than be comprehensive. However, it does not avoid the introduction of difficult ideas and concepts. Indeed, it has been written to draw attention to the relationship of this challenging subject with the more rigorous physical sciences and the existence of deeper analytical principles that both merit and require thoughtful reflection. As it takes the reader through glimpses of the body's systems and how they interact with each other and the world outside, it provokes interest, debate and curiosity about a subject that lies at the very foundation of medicine. If it persuades some of our most talented students that not all their studies in medicine or biology simply involve rote learning, this little volume will have served its purpose. I have every confidence it will do so, and welcome it wholeheartedly.

Christopher Huang
Professor of Cell Physiology
Professorial Fellow of New Hall
University of Cambridge
July 2007

Acknowledgements

Those to whom we owe a debt of gratitude are too numerous to list. However, we are especially grateful to our lecturers and supervisors at the University of Cambridge who inspired our interest in the subject. Our particular thanks go to Dr Richard Barnes, Dr Richard Dyball, the late Dr Anthony Edwards, Dr Stephanie Ellington, Dr Roger Tapp and Dr Teresa Tiffert.

We are also grateful to the medical, veterinary and Natural Sciences students whom we have taught over the past five years. It is their humour, patience and questions that have made teaching such a pleasure and enabled us to foster and maintain our enthusiasm. We are particularly indebted to Gareth Matthews, Angharad Wheeler, Zaina Zafarulla and Dr Victoria Jones for their helpful comments on the manuscript and Dr Thomas Cass for his assistance with the mathematical concepts.

Most of all, we are thankful to Professor Christopher Huang for his seemingly endless support and encouragement for all our endeavours.

Contents

Introduction

Physiology is the study of the mechanical, physical and biochemical functions of living organisms. All living organisms require a stable internal environment for survival. A large part of physiology is therefore about how we achieve this. In the absence of external challenges, this requires simply keeping the internal environment the same. In the face of challenges from the world outside, however, it requires the ability to resist change and respond in ways that tend to bring the state of the system back to normal. Most of physiology is therefore about *minimizing* the effect of change rather than *preventing* change itself.

This distinction is an important one. Each component of our internal environment interacts with innumerable others in innumerable ways. A change in one component will therefore have knock-on effects on others and controlling one is often only possible at the expense of another. Countering the effect of a change in one variable often requires changing another. If the O_2 content of the air we breathe falls, should we make do with less and keep our CO_2 levels normal or breath faster to inhale more O_2 and in so doing, blow off CO_2? If we are short of NaCl, should we loose water and keep the concentration constant or conserve water to maintain volume but allow the concentration to change? If an organ requires an increase in blood flow, should we allow this increase at the expense of blood flow to other organs? In each case, the answer depends on which option poses a greater threat to the internal environment: it will often depend on both the severity and time course of the challenge.

Internal mechanisms are often able to minimise the impact of a change on the body as a whole. However, returning the system to its previous state usually requires alterations in the input or output of substances to or from the body. The impact of the loss of NaCl may be minimised in the short term by a range of mechanisms, but the lost NaCl must ultimately be replaced if the previous state of the body is to be restored. One of the biggest challenges in physiology is therefore balancing input and output.

How such control is achieved can be appreciated by considering the body as a generic *control system*. Information obtained from sensors (*receptors*) is transmitted to a control centre (*controller*), which is charged with the task of deciding what to do. Having made a decision, the controller then transmits a signal to some structure, or structures (*effectors*), thus directly or indirectly effecting a change in the variable. This change may involve using *feedback* or *feedforward*, separately or in combination (Fig. 1). In a *negative feedback system*, an effort is made to return the variable to its normal value. In a feedforward system, the body uses a variety of inputs to anticipate change, hoping to respond to it before it happens. Such control systems are truly fundamental to physiology and are discussed in more depth in the Appendix.

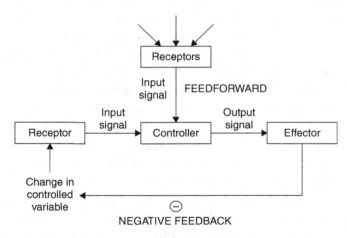

Fig. 1. Feedforward and negative feedback systems.

Appreciating which variables are most important to control and a knowledge of the factors, both internal and external, which influence these are therefore at the core of physiology. This book concentrates on these basic principles. Each of the major systems in the body is covered in turn followed by examples of how these systems interact to face common day-to-day challenges. Further Thoughts sections including additional material follows most chapters and an Appendix is provided at the end for those interested in the theoretical aspects of the subject.

We do not intend this to be another complete textbook of physiology: several excellent comprehensive texts already exist. Instead, we hope to stimulate interest, reflection and discussion and to encourage further reading. If this book conveys even a small part of our great enthusiasm for this fascinating subject, we will have achieved our purpose.

A Note on Units and Symbols

We use standard scientific *Système International* (SI) units for the calculations. This system consists of the following *base units*:

ampere, A: electric current
kelvin, K: temperature
kilogram, kg: mass
metre, m: distance
mole, mol: amount of substance
second, s: time
candela, cd: luminous intensity (not used in this book)

All other units can then be expressed in terms of these base units as *derived units*. The named units used in this book are as follows:

coulomb, C (A·s):   charge
volt, V ($m^2 \cdot kg \cdot s^{-3} \cdot A^{-1}$):   potential difference
farad, F ($C \cdot V^{-1}$, $m^{-2} \cdot kg^{-1} \cdot s^4 \cdot A^2$):   capacitance
ohm, Ω ($V \cdot A^{-1}$, $m^2 \cdot kg \cdot s^{-3} \cdot A^{-2}$):   resistance to electric current
siemen, S ($V^{-1} \cdot A$, $m^{-2} \cdot kg^{-1} \cdot s^3 \cdot A^2$):   conductance to electric current
newton, N ($m \cdot kg \cdot s^{-2}$):   force
joule, J (N·m, $m^2 \cdot kg \cdot s^{-2}$):   energy
watt, W ($J \cdot s^{-1}$, $m^2 \cdot kg \cdot s^{-3}$):   power
pascal, Pa ($N \cdot m^{-2}$, $m^{-1} \cdot kg \cdot s^{-2}$):   pressure

However, in both physiology and medicine, units that are not part of the SI system are in common use. Those appearing in this book are as follows:

centimeter of water, cmH_2O (1 $cmH_2O \simeq 98\,Pa$):   pressure
millimeter of mercury, mmHg (1 mmHg $\simeq 133\,Pa$):   pressure
dalton, Da (one twelfth the mass of a ^{12}C atom, $\simeq 1.66 \cdot 10^{-27}\,kg$):
   atomic mass
litre, L (1 L $= 10^{-3}\,m^3$):   volume

Non-SI units used for the measurement of pressure are explained in the Appendix. M is used to represent $mol \cdot L^{-1}$ and square brackets are used to denote concentrations along with the following subscripts:

[ ]$_e$ extracellular concentration
[ ]$_i$ intracellular or inside concentration
[ ]$_o$ outside concentration
[ ]$_{plasma}$ plasma concentration

The following abbreviations are used to denote powers of ten:

nano, n: 10^{-9}
micro, μ: 10^{-6}
centi, c: 10^{-2}
milli, m: 10^{-3}

Normal Values

$[K^+]_i$	140 mM
$[K^+]_o$	5 mM
$[Na^+]_i$	10 mM
$[Na^+]_o$	140 mM
Resting nerve membrane potential (E_m)	-79 mV
Equilibrium potential for K^+ (E_K) in nerve	-73 mV
Equilibrium potential for Na^+ (E_{Na}) in nerve	$+93$ mV
Nerve membrane capacitance (C_m)	$1\,\mu F \cdot cm^{-2}$
Intra-alveolar pressures during quiet breathing	$-1\,cmH_2O$ to $1\,cmH_2O$

Average intrapleural pressures during quiet breathing	$-5\,cmH_2O$ to $8\,cmH_2O$
Partial pressure of O_2 in arterial blood (P_{aO_2})	$95\,mmHg$
Partial pressure of O_2 in alveolar gas (P_{AO_2})	$100\,mmHg$
Rate of O_2 consumption (V_{O_2}) at rest	$300\,ml \cdot min^{-1}$
Rate of O_2 consumption (V_{O_2}) in extreme exercise	$3\,L \cdot min^{-1}$
Rate of CO_2 production (V_{CO_2}) at rest	$200\,ml \cdot min^{-1}$
O_2 content of arterial blood	$200\,ml \cdot L^{-1}$
Atmospheric pressure (P_{atmos}) at sea level	$760\,mmHg$
Saturated vapour pressure of water at body temperature	$47\,mmHg$
Systolic arterial blood pressure (ABP)	$120\,mmHg$
Diastolic ABP	$80\,mmHg$
Mean ABP ($\overline{ABP}$)	$95\,mmHg$
Mean systemic filling pressure (MSFP)	$7\,mmHg$
Jugular venous pressure (JVP)	$7\,cmH_2O$
Cardiac output (CO) at rest	$5\,L \cdot min^{-1}$
Venous return (VR) at rest	$5\,L \cdot min^{-1}$
Right atrial pressure (RAP)	$0\,mmHg$
Glomerular filtration rate (GFR)	$125\,ml \cdot min^{-1}$
Renal blood flow (RBF)	$625\,ml \cdot min^{-1}$
Plasma osmolarity	$286\,mOsM$
Plasma pH range	7.34–7.44

CHAPTER 1

Electrical Properties of Cells

All multicellular life forms face the complex problem of ensuring that the behaviour of each cell supports the continued function of the entire organism. To this end, intercellular (cell to cell) communication is of paramount importance. Animals have evolved two interacting strategies: chemical and electrical signalling. Chemical signals, secreted by one cell and acting on another, may act locally on neighbouring cells (*paracrine factors*) or be transported in the bulk circulation to act on distant cells (*endocrine factors* or *hormones*). An advantage of this strategy is that signals can be sent to large numbers of cells at the same time. Furthermore, by expressing different receptors, individual cells can respond differently to the same chemical signals. The main disadvantage of chemical signals is that they take a rather long time to act, but this is useful where sustained responses are required. In contrast, electrical signals are highly specific to their target cells and act rapidly to bring about transient responses. There are additionally important interactions between these two modes of signalling: electrical signals frequently trigger chemical signals, which may then go on to trigger further electrical signals.

This chapter focuses on electrical signalling and discusses how these signals are set up and transmitted around the body. Such signalling requires neurons to act as electrical conducting wires, passing current longitudinally between two points. Since a current is a flow of charge, it must be driven by an energy difference (*potential difference*) between the two points. Such potential differences exist transversely across the membranes of all cells at rest (*resting potentials*). However, electrical signalling requires that the cells involved be capable of altering this

1

potential difference, usually moving from being inside-negative at rest to inside-positive, as a result of the movement of charge across their membranes. This change in potential difference is described as *depolarisation* and the ability to depolarise defines *excitable cells* such as nerve and muscle. Here, we first consider how the resting potential arises and then go on to describe the process of depolarisation and discuss how the signals set up are transmitted around the body.

1.1　The Resting Membrane Potential

To understand how a potential difference arises across a cell membrane at rest we only need to think of a cell in very simple terms. We can begin by imagining it as nothing more than a bag of protein, which we will call A. The cell membrane must be impermeable to A so that it does not leak out of the cell. Like other proteins and buffers, A is capable of gaining and loosing H^+ ions. The pK_a of A, the pH at which half of A is capable of gaining H^+ and half is capable of loosing H^+ (see Chapter 7), is around 6.2. Intracellular pH is around 7.2 and at this higher (more alkaline) pH, A will tend to loose H^+ and become negatively charged. In fact, at pH 7.2, A has a mean charge of -1.2 ($A^{-1.2}$). It is because of this charge on A that a potential difference arises across cell membranes at rest (commonly referred to as the *resting potential, resting E_m*). To understand why, we can imagine a situation where the intracellular fluid (ICF) contains only $A^{-1.2}$ and water while the extracellular fluid (ECF) contains only K^+ and water (Fig. 2).

At rest, cell membranes can be assumed to be freely permeable to K^+. From the point of view of diffusion, the lowest energy state will therefore be reached when the concentration of K^+ is the same on both the inside ($[K^+]_i$) and outside ($[K^+]_o$) of the membrane, i.e.

$$[K^+]_i = [K^+]_o$$

However, from the point of view of electric charge, the lowest energy state will be reached when the charge inside the cell equals the

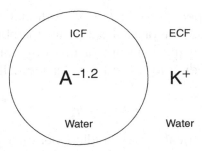

Fig. 2. Simple model cell.

charge outside the cell. Remembering the charge on A, this will occur when:

$$[K^+]_i = [K^+]_o + 1.2[A^{-1.2}]_i$$

Clearly both the above conditions cannot be met at the same time as $[K^+]_i$ cannot both be the same as $[K^+]_o$ and greater than $[K^+]_o$. At equilibrium, $[K^+]_i$ will therefore lie somewhere between these two values. At this equilibrium, there will still be an energy gradient for K^+ to leave the cell from a diffusive (concentration) point of view and an energy gradient for K^+ to enter the cell from an electrical point of view, but these two will balance. It is this persistent electrical gradient, arising from the difference in charge between one side of the membrane and the other, that gives rise to the resting E_m. By equating these two energy gradients, we can obtain an equation for the resting E_m.

For each mole of K^+, the diffusive energy gradient tending to make K^+ leave the cell at equilibrium is given by:

$$\Delta\mu_d = RT \ln \left(\frac{[K^+]_i}{[K^+]_o} \right)$$

where $\Delta\mu_d$ = diffusive energy gradient per mole of K^+ ($J \cdot mol^{-1}$)
R = ideal gas constant ($6.31\ J \cdot mol^{-1} \cdot K^{-1}$)
T = absolute temperature (K)

Thus the energy per mole of K^+ is proportional to the logarithm of relative K^+ concentrations on each side of the membrane, with R and T together constituting the constant of proportionality. The electrical energy gradient tending to make K^+ enter the cell at equilibrium is given by:

$$\Delta\mu_e = zFE_K$$

where $\Delta\mu_e$ = electrical energy gradient per mole of K^+ ($J \cdot mol^{-1}$)
         z = charge on K^+ (its valency, +1)
         F = Faraday constant ($96500\ C \cdot mol^{-1}$)
         E_K = equilibrium potential difference for K^+ (V)

The reasoning underlying this equation is rather straightforward. Here the cell membrane is in effect being treated as a *capacitor*, a device which stores charge. As we show in the Appendix, the energy stored by a capacitor depends on the difference in charge (given by zF for each mole of ions) and the potential difference across it (E_K).

At equilibrium, there is no *net* energy gradient for K^+ movement so the two energy differences must balance. This means that:

$$\Delta\mu_d = -\Delta\mu_e$$

and

$$-RT \ln \left(\frac{[K^+]_i}{[K^+]_o} \right) = zFE_K$$

Since

$$-\ln a = \ln \left(\frac{1}{a} \right)$$

this can be rearranged to

$$E_K = \frac{RT}{zF} \ln \left(\frac{[K^+]_o}{[K^+]_i} \right)$$

which is the Nernst equation. This gives the equilibrium (Nernst) potential difference across a membrane which is permeable only to one ion, in this case K^+. Note that this equilibrium potential depends only on the relationship between $[K^+]_i$ and $[K^+]_o$. Providing the membrane remains permeable to K^+, altering the permeability does not alter the equilibrium potential.

Putting standard values of R (6.31 J·mol^{-1}·K^{-1}), T (298K), z (+1) and F (96500 C·mol^{-1}) and converting from ln to log$_{10}$ gives:

$$E_K = 58\,\text{mV}\,\log_{10}\left(\frac{[\text{K}^+]_o}{[\text{K}^+]_i}\right).$$

Unsurprisingly, real cells are not quite so simple. Cell membranes are permeable not only to K$^+$ but also to other ions. Of these, only Na$^+$ makes a significant contribution to the resting E_m. We can see the effect this has by plotting experimentally measured values of resting E_m against $[\text{K}^+]_o$ at two different values of $[\text{Na}^+]_o$ (Fig. 3).

Even when $[\text{Na}^+]_o$ is at its physiological value in humans (140 mM, broken line) the curve deviates from the values of E_K calculated using the equation above, especially when $[\text{K}^+]_o$ is low. When $[\text{Na}^+]_o$ is raised to 1000 mM, there is a large difference between the observed curve and that of which the Nernst equation predicts.

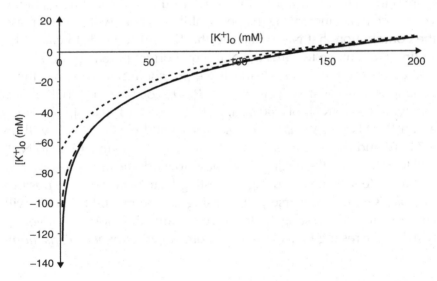

Fig. 3. Relationship between E_m and $[\text{K}^+]_o$. Solid line is the prediction from the Nernst equation, broken line shows measured values at physiological levels of $[\text{Na}^+]_o$ and dotted line shows measured values at $[\text{Na}^+]_o = 1000$ mM.

We can account for this contribution from Na^+ in a modified form of the Nernst equation, the Goldman equation

$$E_m = \frac{RT}{zF} \ln \left(\frac{P_K[K^+]_o + P_{Na}[Na^+]_o}{P_K[K^+]_i + P_{Na}[Na^+]_i} \right),$$

where P_{Na} is the permeability of the membrane to Na^+ and P_K the permeability of the membrane to K^+.

From this, we can see that the resting E_m across a membrane that is permeable to more than one ion will lie somewhere between the equilibrium potentials for the individual ions, weighted by the relative permeabilities to those ions. At rest, a cell membrane is much more permeable to K^+ than it is to Na^+ (by a ratio of approximately 100:1). Na^+ is therefore excluded from the cell and $[Na^+]_o$ is larger than $[Na^+]_i$. This results in a positive Na^+ equilibrium potential, E_{Na}. In contrast, the high K^+-permeability of the membrane allows the negative change on $A^{-1.2}$ to pull K^+ into the cell, resulting in $[K^+]_i$ being larger than $[K^+]_o$. This results in a negative E_K. The fact that the membrane is far more permeable to K^+ results in the resting E_m being very close to E_K. Increasing the permeability to Na^+ will then increase the contribution that Na^+ makes to the E_m and so tends to cause the resting E_m to move towards the Na^+ equilibrium potential, E_{Na}.

We can see this more clearly if we put some numbers into these equations. In a resting cell with a P_{Na}/P_K of 0.01 at 310 K (body temperature), a $[K^+]_i$ of 140 mM, a $[K^+]_o$ of 5 mM, $[Na^+]_i$ of 10 mM and a $[Na^+]_o$ of 140 mM, the resting E_m will be -79 mV, E_K will be -73 mV and E_{Na} will be $+93$ mV. Increasing the permeability to Na^+ will tend to cause the resting E_m to move towards the more *positive* E_{Na} as Na^+ enters the cell, causing the resting E_m to become more *positive* (*depolarisation*). Conversely, increasing the permeability to K^+ will tend to cause the resting E_m to move towards the more *negative* E_K, and thus the resting E_m will become more *negative* (*hyperpolarisation*).

1.2   Ion Channels

We can now move on to see how changes in E_m can be brought about by changes in the relative permeabilities of the cell membrane to Na^+

and K^+. In this context, *permeability* refers to the degree to which the plasma membrane allows a particular substance to cross. This is dealt with more fully in Further Thoughts. Since small ions such as Na^+ and K^+ have too high a charge density to pass through the cell membrane itself, they require specialised ion channels in order to cross. The permeability of the membrane to these ions is directly related to the number of open ion channels. Each single ion channel can then be thought of as a resistor. The size and direction of the current flowing through such a resistor depends only on its resistance (or conductance, the reciprocal of resistance) and the potential difference across it.

The relationship between current and potential difference for a single voltage-activated Na^+ channel is show in Fig. 4. This relationship is a straight line: Na^+ will tend to enter when the E_m is less than the E_{Na} (inward Na^+ current) and leave when the E_m is greater than E_{Na} (outward Na^+ current). The E_m at which the net direction of ion movement across the membrane (net direction of the current) switches is referred to as the *reversal potential*. Since this particular ion channel is permeable only to Na^+, its reversal potential is E_{Na}. If instead, the channel is permeable to both Na^+ and K^+ (Fig. 5), then at all E_ms between E_K and E_{Na}, K^+ will tend to leave the cell (outward K^+ current)

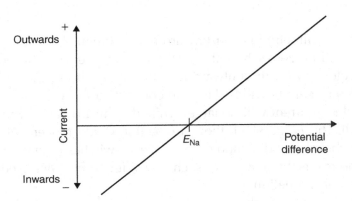

Fig. 4. The relationship between current and potential difference for a single voltage-activated Na^+ channel.

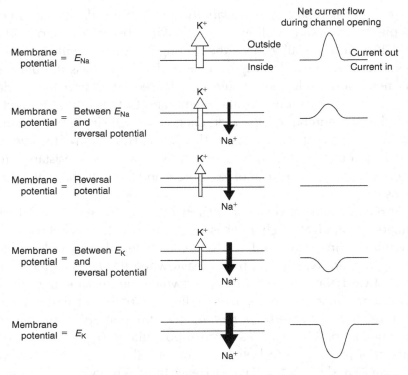

Fig. 5. Current flow and reversal potential of a channel permeable to Na^+ and K^+.

while Na^+ will tend to enter (inward Na^+ current). At E_K, the net K^+ current will be zero while at E_{Na}, the net Na^+ current will be zero. At E_ms close to E_K, the outward K^+ current will be small, while the inward Na^+ current will be large. In contrast, at E_ms close to E_{Na}, the outward K^+ current will be large, while the inward Na^+ current will be small. The E_m at which the net direction of ion movement across the membrane (net direction of the current) switches is again referred to as the reversal potential. In such cases, it can be calculated using the Goldman equation.

Interestingly, but perhaps not altogether surprisingly, not all current-voltage relationships are as simple as that of the voltage-activated

The Na$^+$/K$^+$-ATPase

Since a potential difference between two points arises as a result of a charge difference between those points, anything that contributes to the difference in charge across the membrane will contribute to the E_m. The Na$^+$/K$^+$-ATPase is a particularly important example. This protein is found in the membrane of every cell. It usually transports 3 Na$^+$ out of the cell and brings in 2 K$^+$ in exchange, consuming ATP in the process. Such transporters that directly consume ATP are referred to as *pumps*, while those that do not are referred to as *exchangers*. As a result, each pump cycle leads to the loss of one unit of positive charge from the cell. This loss of positive charge results in the ICF becoming even more negative with respect to the ECF, and therefore the resting E_m becomes more negative, i.e. the pump is *electrogenic*. The effect of this pump can be incorporated into a calculation of resting E_m with a modification of the Goldman equation, called the Mullins-Noda equation:

$$E_m = \frac{RT}{zF}\ln\left(\frac{r[K^+]_o + \frac{P_{Na}}{P_K}[Na^+]_o}{r[K^+]_i + \frac{P_{Na}}{P_K}[Na^+]_i}\right)$$

Here r represents the number of Na$^+$ ions pumped out for each K$^+$ ion that enters the cell on the Na$^+$/K$^+$-ATPase, i.e. in most cells $r = 1.5$. However, a simple calculation shows that in most cell types, the Na$^+$/K$^+$ATPase contributes only around $-6\,$mV to the resting E_m. The Na$^+$/K$^+$ATPase therefore plays *only a negligible part* in establishing the resting E_m. This pump is vitally important however its key role is in the maintenance of cell volume. This is discussed in detail in Further Thoughts.

Na$^+$ channel: some ion channels allow the passage of current in one direction only, in effect acting as diodes (Fig. 6). Such channels are described as *rectifiers* and one of the best examples is the inwardly rectifying K$^+$ channel found in the ventricles of the heart.

In the case of voltage-activated ion channels, the probability of an individual channel being open increases as the membrane depolarises.

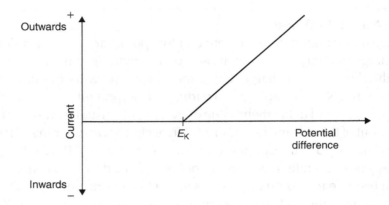

Fig. 6. The relationship between current and potential difference for a single rectifying K^+ channel that only allows current to flow out of the cell.

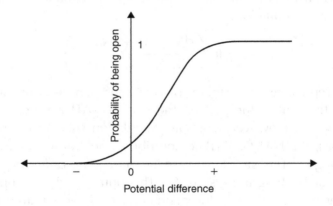

Fig. 7. Na^+ channel activation curve.

The voltage-activated Na^+ channels, found in the membranes of excitable cells, constitute a particularly important class of such channels. The relationship between E_m and the probability that a single Na^+ channel will be open, referred to as an *activation curve*, is shown in Fig. 7. It can be seen that the relationship is both non-linear and increasing (it is termed a *Boltzmann curve*). Note also that because these channels open in response to depolarisation and an increase

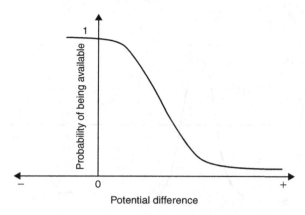

Fig. 8. Na$^+$ channel inactivation curve.

in Na$^+$ permeability itself causes depolarisation, any depolarisation will result in a positive feedback effect. As more Na$^+$ channels open, the inward Na$^+$ current increases and this further depolarises the E_m, opening more Na$^+$ channels. By contrast, the opening of voltage-activated K$^+$ channels in response to depolarisation results in *hyperpolarisation* (recall that E_K is more negative than the resting E_m) which forms a negative feedback effect bringing the E_m back towards its resting value. We will see that both these properties are central to the generation of an action potential.

As well as showing this voltage-dependent activation, these Na$^+$ channels also inactivate and close in a voltage-dependent manner. The relationship between E_m and this inactivation is described by an *inactivation curve*, shown in Fig. 8 (again a Boltzmann curve). This shows that the channels also *close* during membrane depolarisation. However, inactivation is slower than activation so the overall effect of membrane depolarisation is to cause Na$^+$ permeability to rise rapidly to a peak before falling again.

1.3 The Nerve Action Potential

Having shown that the resting E_m is close to the equilibrium potential for K$^+$ and that transient changes in the E_m result from changes in the permeability of the membrane to Na$^+$ and K$^+$, we can now begin

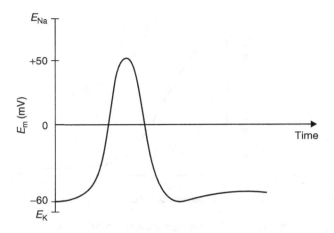

Fig. 9. A nerve action potential.

to understand how a nerve action potential (Fig. 9) is generated. The story of how this point of understanding was reached is fascinating. The experimental techniques that were used were truly innovative and these are given some attention in Further Thoughts.

All action potentials begin from a small depolarisation, either arising at a receptor or in another part of the nerve. This small depolarisation results in the opening of voltage-activated Na^+ channels which increase the Na^+ permeability of the membrane, leading to further depolarisation and the opening of more voltage-activated Na^+ channels and so on. This positive feedback continues until these channels inactivate, causing Na^+ permeability, and hence depolarisation, to reach a peak and then begin to fall. At around that time voltage-activated K^+ channels begin to open (they open in response to depolarisation but only after a short delay) and rapidly pull the E_m back down towards E_K. Unlike the Na^+ channels, these K^+ channels do not inactivate during the time course of an action potential so after a slight overshoot, the negative feedback restores the resting E_m.

Several key observations of action potentials can be explained with reference to the properties of the ion channels described above. First, the notion of *threshold*. The threshold is the minimum change in

Clinical Box 1: *Mutations in Voltage Activated Na$^+$ Channel Genes*
Mutations in voltage-activated Na$^+$ channel genes which either enhance activation or retard inactivation underlie a number of clinical disorders. A rightward shift in the inactivation curve results in retarded inactivation, sustained depolarisation and continuous contraction in certain cases of persistent severe muscle stiffness *(myotonias)* and paralysis in humans. Interestingly, selective breeding has resulted in such mutations being especially common in thoroughbred horses *(equine periodic paralysis)*. Affected animals have highly developed muscles, perhaps as a result of myotonic contractions which are tantamount to continuous exercise! Mutations resulting in retarded Na$^+$ channel inactivation also underlie one form of the *Romano-Ward syndrome*, a rare but interesting cause of abnormal cardiac rhythms *(arrhythmias)* which can result in sudden death. In this case, failure of Na$^+$ channel inactivation increases the inward depolarising current during the cardiac action potential. This ultimately results in arrhythmia, although the intervening mechanism remains unclear. Part of the story is probably that the prolonged action potential duration results in the membrane being sufficiently depolarised to activate voltage-activated Ca^{2+} channels after they have recovered from inactivation, resulting in abnormal extra action potentials. More rarely, a leftward shift of the activation curve results in enhanced activation, depolarisation and sustained contraction and has the same effect.

the E_m needed to initiate an action potential. It represents the E_m at which the number of Na$^+$ channels open allows the inward Na$^+$ current to exceed the outward K$^+$ and result in a positive-feedback depolarisation. As a result of this positive-feedback and the subsequent inactivation of Na$^+$ channels, a depolarisation of a magnitude greater than this threshold will initiate an action potential of a standard amplitude and duration. A subthreshold stimulus will not result in an action potential at all. This is described as the *all-or-nothing law*: the magnitude and form of the action potential are not dependent on the strength of the stimulus.

Na^+ Entry During an Action Potential

It is often said that Na^+ "floods" across the membrane during an action potential. If a cell membrane depolarises by 100 mV during an action potential and the capacitance (the ability to store charge) of each square centimetre of membrane is around 1 μF, then the amount of charge which must cross the membrane to achieve this depolarisation can be calculated as:

$$\Delta Q = C\Delta V$$

where Q = charge (C)
C = capacitance (C $\cdot$ V^{-1}, F)

$$\Delta Q = 10^{-6} \cdot 100 \cdot 10^{-3} = 10^{-7} C$$

Each mole of univalent ions carries a charge of around $10^5 C$ (Faraday's constant) and therefore the amount of Na^+ crossing each square centimetre of membrane during an action potential is: $\frac{10^{-7}}{10^5} = 10^{-12}$ mol or 1 pmol — hardly a "flood".

The small membrane capacitance and large value of Faraday's constant together mean that only miniscule differences in the concentrations of anions and cations between one side of the membrane and the other are needed to produce physiological membrane potentials. In fact, is it often assumed that these concentrations are identical on each side of the membrane (the *principle of electroneutrality*). Of course, this is only *approximately* true.

Secondly, the notion of the *relative refractory period*. This is the time following an action potential when the threshold is increased and a greater depolarisation is required to initiate an action potential. This arises because depolarisation renders a proportion of Na^+ channels inactivated (Fig. 8), reducing the absolute number of Na^+ channels available for activation. However, the same critical number of Na^+ channels must still be activated in order to produce a sufficient inward Na^+ current to overcome the outward K^+ current and initiate an action potential. A larger depolarisation (step to the right shift in Fig. 7) is therefore required to activate a large enough proportion of the remaining Na^+ channels available for activation. Furthermore, the time following an action potential when so many Na^+ channels are inactivated that there are not enough left to elicit a second action potential is referred to as the *absolute refractory period*. While theoretically important, this is a practically less useful concept as measuring it would require an infinitely large stimulus amplitude.

Thirdly, the phenomenon of *accommodation*. This refers to an increase in threshold following prolonged depolarisation and a decrease in threshold following prolonged hyperpolarisation. As we saw in the Na^+-channel inactivation curve (Fig. 8), Na^+-channels inactivate as the membrane depolarises, decreasing the number available for activation. These channels then remain inactivated until the E_m repolarises. Prolonged depolarisation therefore results in a prolonged decrease in the proportion of Na^+-channels available to open which increases threshold as explained above. Conversely, hyperpolarisation of the membrane to values more negative than the resting E_m decreases the proportion of inactivated Na^+-channels and so lowers the threshold required to initiate a second action potential.

Action Potentials in Other Excitable Cells

The description so far particularly reflects the events occurring during a nerve action potential. Other excitable cells express other ion channels which modify the form and duration of the action potential. For example, cells of the heart also express voltage-activated Ca^{2+} channels, opened as a result of depolarisation attributable to voltage-activated Na^+ channels. A large transmembrane gradient in $[Ca^{2+}]$ exists (10^4 times higher in the ECF as compared to the ICF) and therefore Ca^{2+} flows inwards on channel opening, prolonging the depolarisation. As will be seen in the next chapter, this influx of Ca^{2+} is also critical in the initiation of contraction.

1.4 Action Potential Propagation

Having understood the ionic basis underlying a single action potential, we can now think about how these electrical signals are propagated along nerve axons. The propagation of an action potential along an unmyelinated (see later) nerve axon can be modeled as the longitudinal spread of Na^+ current along an infinitely long insulated cable. The nerve axon is then represented as a network of resistors and capacitors through which this current must flow (Fig. 10).

Most undergraduate textbooks then go on to make the implicit assumption that an action potential is an infinitely long constant current applied for an infinitely long time transversely across the centre of the cable. This simplifies matters a great deal. If this assumption is made, then moving away from this, the magnitude of the current successively halves over a constant distance, i.e. it decays exponentially with distance. This leads to the notion of a space constant, a simple parameter describing the decay of current with distance, and some apparently very straightforward ideas about conduction velocity. These ideas are dealt with more thoroughly in Further Thoughts.

However, this simplification is *not* valid. Action potentials are *not* infinitely long events but rather are brief, only a few milliseconds duration in nerves. During this time, the potential difference across the membrane changes rapidly. This is important. As we have established,

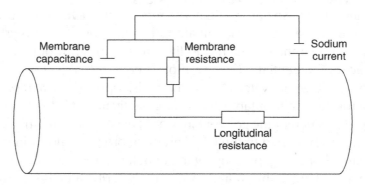

Fig. 10. Circuit model of a nerve axon.

the cell membrane can be modeled as consisting of both a resistor and a capacitor. While the current through a resistor depends on the potential difference across it, the current through a capacitor depends on the *rate of change* of the potential difference across it. This is explained in the Appendix. The rapid change in potential difference during an action potential makes the current through the capacitor far larger than the current through the resistor. This makes the capacitance of the cell membrane of great importance, and its resistance of little importance, in determining how much current is wasted before it can begin to spread longitudinally. The smaller this capacitance, the less current will be consumed in discharging it and the further it will spread. The membrane capacitance can be made smaller by increasing the distance between the capacitor plates, as explained in the Appendix. This is achieved by wrapping the axon with an insulating layer of *myelin*.

It follows that the longitudinal resistance of the axon must be key in determining how far along the axon the current will spread. The thicker the axon, and therefore the lower this resistance (see Appendix), the further along the axon the current will spread. One final factor must affect this distance, the size of the inward Na^+ current across the membrane driving these events. The larger this value, the further the current will propagate longitudinally. Increasing Na^+ channel density therefore increases this distance.

Importantly, the magnitude of the current generated by an action potential decays below the value needed to cause the E_m to reach threshold over a distance of a few millimetres. If information only needs to be transmitted over these very short distances, as in the eye, then such passive longitudinal spread of current is all that is needed. However, this is nowhere near long enough to allow the current resulting from a single action potential to transmit information large distances around the body. To achieve this goal, multiple action potentials must be actively regenerated. During an action potential, a longitudinal potential difference arises between the depolarised region of the membrane and adjacent areas still at E_m. Current therefore then flows longitudinally inside the axon from depolarised (positive) to resting (negative) and outside the axon from resting (positive) to

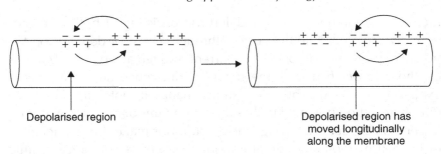

Depolarised region   Depolarised region has
                        moved longitudinally
                        along the membrane

Fig. 11. Longitudinal spread of depolarised region along an axon.

depolarised (negative), setting up an electrical circuit (Fig. 11). This current depolarises the adjacent membrane, setting up a new action potential. In this way, once an action potential has been initiated at one point it will be propagated along the entire length of the nerve axon.

1.4.1 *Propagation along myelinated nerves*

A proportion of the nerve axons in the body are wrapped with an insulating layer of myelin. In the case of unmyelinated axons, the above discussion is adequate. However, interruptions in this myelin sheath (*nodes of Ranvier*), which occur approximately every milimetre, make propagation along myelinated axons somewhat different (Fig. 12).

In the *internodal regions*, this wrapping of myelin prevents the nerve cell membrane from coming into contact with the ECF. This prevents Na^+ ions from crossing the membrane and therefore prevents action potentials from being regenerated. However, at the nodes, the absence of myelin allows such regeneration. Notably Na^+-channels cluster in these regions. A given depolarisation will therefore result in a far larger inward sodium current at the nodes than at the internodes. In practice, this results in the internodes behaving as passive cables: the purpose of these connecting regions is simply to deliver current to the next

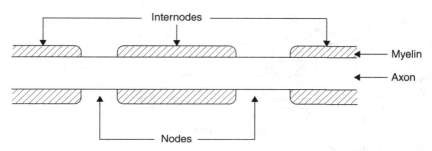

Fig. 12. A myelinated nerve showing nodes and internodes.

Clinical Box 2: *Multiple Sclerosis*

In multiple sclerosis, a disease of the central nervous system, myelin sheaths degenerate resulting in a decrease in the membrane resistance. This results in an increase in the rate at which the magnitude of a depolarisation decays with distance. Clearly, if the magnitude of a depolarisation decays to below threshold before the next internode is reached, propagation of the action potential will cease. Fortunately, while the internodal distance is around 1 mm, the potential difference is usually still large enough to initiate a new action potential 2 mm to 4 mm along the axon. This provides a generous *safety factor*: ordinarily, a depolarisation will skip several nodes and if the distance over which E_m is suprathreshold is reduced there is considerable reserve available. A further reduction in this distance may explain *Uhthoff's phenomenon*, previously used as a diagnostic test for multiple sclerosis. This describes a worsening of symptoms when body temperature is raised, for example after taking a hot bath. As temperature increases, K^+ conductance increases to a greater degree than Na^+ conductance and therefore threshold increases. If the E_m is then subthreshold at the first internode, the action potential is lost. This is illustrated in Fig. 13.

---

node at a sufficient magnitude to initiate another action potential, i.e. above threshold. This discontinuous ("jumping") conduction along a myelinated nerve axon is referred to as *salutatory conduction*.

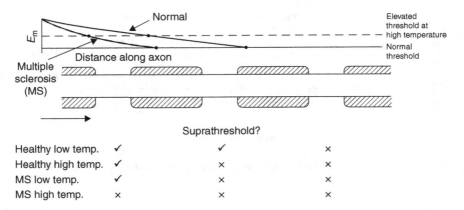

Fig. 13. Action potential propagation in normal individuals and in sufferers of multiple sclerosis.

   While the passive spread of current is fast, the regeneration of action potentials is slow. This is the key reason why myelinated axons on the whole allow faster propagation than unmyelinated axons. In the latter, action potentials constantly, and unnecessarily, regenerate.

   The increase in conduction velocity afforded by myelination is essential for survival. If all our nerve axons were unmyelinated, it might take anything up to 10 seconds between deciding to run away from a lion and actually starting to run. However, myelination does not *always* increase conduction velocity: in fact, for very narrow axons (less than around 1.5 μm in diameter) myelination appears to decrease it (Fig. 14).

1.5   Synapses and the Neuromuscular Junction

Having discussed in some detail how signals are carried along nerves, we must now think about how these signals are turned into *useful* responses. In general, nerves may be connected to other nerves or directly to *effector organs*. Such organs are triggered by nerve signals to bring about responses. For example, activity in motor nerves initiates skeletal muscle contraction while activity in nerves supplying the salivary glands brings about secretion.

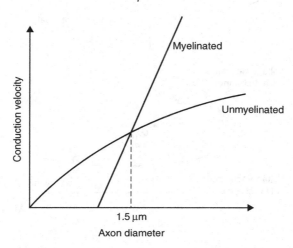

Fig. 14. Relationships between axon diameter and conduction velocity.

Communication between nerves and effector organs (*neurotransmission*) takes place via *synapses*. The same two possible means of intercellular communication discussed previously, namely chemical and electrical, are also available at synapses. Chemical neurotransmission is the most commonly used strategy as it allows greater flexibility and the possibility of inhibitory signalling. This is facilitated by the very small distances (in the micrometer range) across which such signals are sent. The key advantages of electrical neurotransmission are speed and the ability to transmit information in both directions. This makes this strategy well-suited for certain roles in the brain where rapid information transfer is essential, as well as for communication between individual cardiac and smooth muscle cells via *gap junctions*. Here we will briefly consider the mechanism by which chemical neurotransmission takes place, focussing on transmission across the neuromuscular junction as an important and well-studied example.

On arrival of an action potential at the presynaptic terminal, depolarisation results in the opening of voltage-activated Ca^{2+} channels and the influx of Ca^{2+} (Fig. 15). This Ca^{2+} then binds to the surfaces of small sacs of neurotransmitter (*vesicles*) already

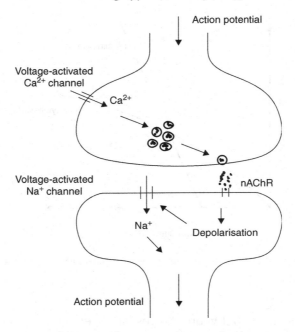

Fig. 15. The synaptic terminal.

present inside the presynaptic terminal. At the neuromuscular junction, this neurotransmitter is acetylcholine (ACh). ACh is synthesised from acetyl co-enzyme A (diverted from the *Krebs cycle*, see Chapter 3) and choline (a form of vitamin B) by the aptly named enzyme choline-acetyltransferase inside the presynaptic terminal. The binding of Ca^{2+} to the surfaces of these vesicles results in their moving towards the far end of the presynaptic terminal, fusing with the plasma membrane and releasing their contents into the *synaptic cleft*. The release of neurotransmitter vesicles is in fact a random process, the probability of which is increased with increasing $[Ca^{2+}]_i$. In general, this process is all-or-none: either the entire contents of a vesicle is released, or none at all. However, a recent study has suggested that this may not always be the case in the brain and that instead, a neurotransmitter-containing vesicle may occasionally fuse with the presynaptic membrane, release some of its contents, then reseal and move away from the membrane, a process referred to in the paper as "kiss-and-run".

The ACh then diffuses across the synaptic cleft and binds to post-synaptic receptors on the muscle cell membrane, before later being broken down by the enzyme acetylcholinesterase. These receptors, termed nicotinic ACh receptors (nAChRs) as they open in response to the experimental drug nicotine, are in fact ion channels. However, rather than opening in response to deplarisation, these *ligand-gated ion channels* open in response to the binding of ACh. The nAChR is the best studied example of such a ligand-gated ion channel and has a number of fascinating molecular and electrophysiological properties, most of which are sadly beyond the scope of this book. The nAChR is permeable to all monovalent cations. Interestingly, since:

$$[Na^+]_i + [K^+]_i \approx [Na^+]_o + [K^+]_o$$

applying the Goldman equation shows that its reversal potential must be close to 0 mV. Opening of thousands of these ion channels results in depolarisation of the post-synaptic cell (an excitatory post-synaptic potential, EPSP, often referred to as an end-plate potential, EPP, at the neuromuscular junction). If the EPSP is sufficiently large, it then goes on to initiate an all-or-nothing muscle action potential, ultimately resulting in muscle contraction.

Chemical synapses at other locations in the body, including the *autonomic nervous system* (ANS), work in similar ways. The ANS is a part of the output division of the peripheral nervous system which innervates, either directly or indirectly, three types of cell — smooth muscle cells, cardiac muscle cells and secretory cells. It is divided into two distinct systems — the *sympathetic nervous system* and the *parasympathetic nervous system*. Each has a wide range of effects but, in general, the sympathetic nervous system can be considered responsible for the "fight, fright and flight" response whilst, the parasympathetic nervous system is more concerned with "rest and digest". These two systems often have opposite, antagonistic, effects on the same organ. Synapses in the autonomic nervous system also make use of interactions between different neurotransmitters, including not only "classical" transmitters such as ACh and *noradrenaline* but also neuropeptides such as *vasoactive intestinal peptide* (VIP) and even gases such as nitric oxide. Contrary to earlier theories, in recent

---

Clinical Box 3: *Myasthenia Gravis*

Myasthenia gravis is a relatively common autoimmune cause of muscle weakness and may be life-threatening. It results from the production of antibodies against post-synaptic nAChR. These block the channels, decreasing the size of EPPs and consequently making it harder to generate muscle action potentials. Myasthenia gravis has classically been treated by giving drugs that block the action of acetycholinesterase, increasing the concentration of ACh in the synaptic cleft, and consequently the proportion of nAChRs to which ACh is bound. Similar effects result from antibodies acting against the presynaptic voltage-activated Na^+ channel in Lambert-Eaton syndrome. Unfortunately, this is not quite so straightforward to treat.

---

years, it has become clear that the interaction of these non-classical neurotransmitters is of great importance. For example, a combination of ACh and VIP is far more effective in producing salivary secretion than ACh alone, even at high concentrations. This is referred to as *synergism*.

It has also become clear that the pattern of activity in presynaptic nerves is of great importance in determining the degrees to which different neurotransmitters are secreted and the ultimate responses of effector organs. It seems that while regular activity favours the secretion of classical neurotransmitters, bursting activity, where an equal number of impulses are delivered during a particular period of time but are closer together (Fig. 16), tends to favour the secretion of neuropeptides and produces far greater effector organ responses. This may be explained by the observation that while classical

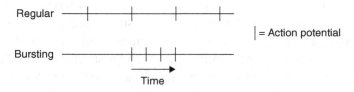

Fig. 16. Regular versus bursting activity in nerves.

neurotransmitters are usually found in vesicles close to the presynaptic terminal, neuropeptide-containing vesicles are usually located deep inside the presynaptic terminal, further away from the membrane at which they are released.

FURTHER THOUGHTS

1.6 A Historical Perspective on the Action Potential

Since Galvani first demonstrated that a frog's leg could be made to twitch by touching it with two different metals, it has been suspected that certain tissues are able to transmit information electrically. It has since become clear that while every cell has a resting E_m, excitable cells including nerve, muscle and even lymphocytes, are able to undergo sudden, transient changes in their E_ms. The first realistic proposal for a mechanism by which E_m might change was made by Bernstein in 1902. Following on from the work of Nernst, Bernstein's K^+ *hypothesis* suggested that while the resting cell membrane was freely permeable to K^+ only, on activation, it became suddenly and transiently freely permeable to all inorganic ions. The sudden relief of osmotic constraints and the resulting influx of positive charge would lead to depolarisation. This results in a readily testable prediction: positive charge should continue to move into the cell until there is no further energy gradient for charge movement, i.e. the E_m is zero. Unfortunately, technical constraints were limiting. In order to measure E_m, the potential difference across the membrane, it is clearly necessary both to know the potentials outside and *inside* the cell, the latter requiring an intracellular electrode. Suitable electrodes and signal amplifiers were not developed for a further two decades and it was not until 1939 that Hodgkin and Huxley published their now classic study, confirming a part of Bernstein's hypothesis but refuting another. While the resting E_m was indeed very close to E_K, during an *action potential*, E_m significantly over-shot zero.

Clearly, there was more to the story. The knowledge that Na^+ was far more abundant in the ICF than in the ECF provided a clue. This leads E_{Na} to be positive and close to the most positive E_m reached during an action potential. This raised suspicions that simple modification of Bernstein's hypothesis could provide the answer. While Cole and Curtis published results indicating that E_m changes during an action potential were unaffected by replacing all ions in the ECF with their osmotic equivalent of glucose, in 1949, Hodgkin and Katz published some quite different results. Their studies clearly indicated that reducing $[Na^+]_o$ reduced the maximum E_m reached during action potential. What is more, the gradient of curve relating the maximum E_m reached during an action potential to $\log_{10}[Na^+]_o$ was almost exactly 58 mV, perfectly in fitting with the prediction of the Nernst equation (Fig. 17).

So how can Cole and Curtis' results be explained? It later turned out that their laboratory technician had been making up their supposedly Na^+-free solutions not using distilled water but instead, using tap water, which of course contains abundant Na^+!

In the following years, many experimenters made great efforts to further elucidate the ionic basis of the action potential in nerve. Experiments by Keynes and others using radioactive isotopes of Na^+ and K^+ provided further important information: a rapid series of action potentials results in a net influx of Na^+ and efflux of K^+. However, the

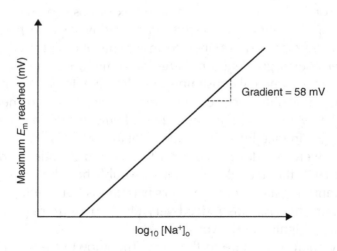

Fig. 17. Relationship between $\log_{10}[Na^+]_o$ and maximum E_m during action potential.

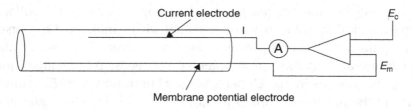

Current electrode

E_c

E_m

Membrane potential electrode

Fig. 18. Voltage clamp set-up.

time resolution of such techniques was limiting and it was not really possible to make a detailed study of ionic movements during a single action potential. Happily, such experiments were soon done. In the early 1950s, Hodgkin, Huxley and Katz published a seminal series of papers describing work using the *voltage clamp technique* (Fig. 18). The idea was simple. During an action potential, two variables change: E_m and time. In order to study mechanisms underlying the change in E_m, it is necessary to fix one variable and since time cannot be fixed, E_m is the obvious choice.

Hodgkin, Huxley and Katz's work relied on a then-recent rediscovery, the squid giant axon. These nerve fibres have diameters of up to 0.5 mm, large enough to allow not one but two microelectrodes to be inserted.

Fundamentally, the ingenuity of this apparatus is that E_m is brought under the experimenter's control. E_c, the *command potential*, is chosen by the experimenter. The *voltage electrodes* measure E_m and the *comparator* (indicated by a triangle on the diagram above) constantly compares it with E_c. If $E_m \neq E_c$, the apparatus injects or removes just enough charge to bring E_m to E_c, i.e. E_m is clamped at E_c. Now:

$$I = \frac{d\Delta Q}{dt}$$

where I = current $(C \cdot s^{-1}, A)$
t = time (s)

Saying that charge, Q, moves across the membrane is therefore equivalent to saying that a current, I, flows. This is measured with an ammeter. We now introduce a *voltage step*: E_c is suddenly changed from its initial value, the *holding potential*, perhaps $-100\,mV$, to a new more positive value, the *test potential*, perhaps −50 mV. If the voltage step exceeds threshold, an action potential will be initiated. Of course, usually during an action potential, an ionic current flows across the membrane and, as the action potential proceeds, ΔQ changes and results in changes in E_m. Under voltage clamp, however, E_m cannot change: the apparatus applies a current which is exactly equal and opposite to that moving across the membrane such that E_m is clamped at E_c. If we can measure the current the apparatus applies, we have in effect measured the current across the membrane (Fig. 19).

There is at least one complication to be dealt with: introducing a voltage step changes E_m. Obviously, in order to change E_m, we must change ΔQ. Now:

$$I = \frac{d\Delta Q}{dt}$$

and

$$\Delta Q = C \Delta V$$

so

$$I = C \frac{dV}{dt}$$

Therefore, during the voltage step, a current will flow as the charge stores by the membrane changes. This *capacitative spike current* is

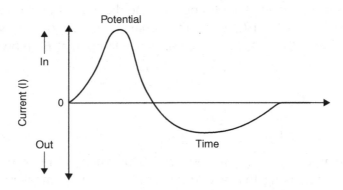

Fig. 19. Current during an action potential.

given by:

$$I_{cap} = C_m \frac{dE_m}{dt}$$

Well designed electronics can reduce this current to zero within $20\,\mu s$, much shorter than the time course of the action potential. Therefore the applied currents truly reflects the transmembrane current.

If we repeat the experiment in the presence of the Na^+-channel blocker tetrodotoxin (TTX, from the ovaries of puffer fish), we obtain the results shown in Fig. 20.

Graphically subtracting the second curve from the first demonstrates that the inward current flowing during the action potential in nerve is attributable to the entry of Na^+ (Fig. 21).

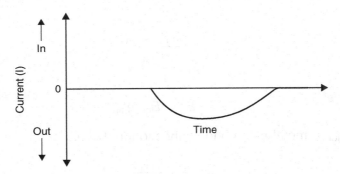

Fig. 20. Current during an action potential in the presence of TTX.

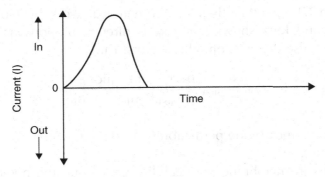

Fig. 21. Na^+ current during an action potential.

Repeating experiment again in the presence of a K^+ channel blocker such as tetraethylammonium (TEA) reveals that the remainder of the later outward current is attributable to K^+. In order to avoid using drugs which might potentially have non-specific actions it is possible to use electrophysiological manoeuvres to separate ionic currents. For example, if $[ion]_i = [ion]_o$, there will be no gradient for flow and therefore the current is eliminated. Further, if $E_c = E_{ion}$, there will be no potential difference driving flow.

Knowing the ionic currents flowing it is a simple matter to convert these to changes in ionic conductance, g_{ion} (the reciprocal of resistance, r_{ion}). By Ohm's law:

$$I = \frac{\Delta V}{R}$$

so

$$I_{ion} = \frac{E_m - E_{ion}}{R_{ion}}$$

or

$$I_{ion} = (E_m - E_{ion})g_{ion},$$

where g_{ion} = membrane conductance to ion (Ω^{-1}).

$$g_{ion} = \frac{I_{ion}}{(E_m - E_{ion})}.$$

It may be more intuitive to consider the permeability of a membrane to an ion (P_{ion}), rather than g_{ion}. Theoretical work by Goldmann, Hodgkin and Katz shows that P_{ion} is directly proportional to g_{ion} although it also depends on other factors. Thus:

$$P_{ion} = g_{ion} \frac{(RT)^2}{F^3 E_{ion}} \frac{[ion]_o [ion]_i}{[ion]_o - [ion]_i},$$

where P_{ion} = membrane permeability to ion ($kg^2 \cdot m^2 \cdot V^{-2} \cdot A^{-2} \cdot s^{-5}$ in this case!).

Changes in membrane permeability underlying the nerve action potential are shown in Fig. 22.

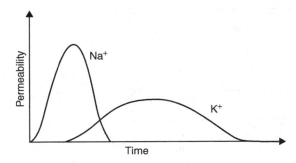

Fig. 22. Changes in membrane permeability to Na^+ and K^+ during an action potential.

1.7 Extracellular Recordings of Action Potentials

To record changes in the E_m during an action potential, a microelectrode must be inserted into a cell and the potential at its tip compared to the potential recorded by an extracellular electrode (Fig. 23). If instead, a nerve action potential is recorded using two extracellular electrodes as in Fig. 24, the potential difference between two points along the axon membrane, rather than E_m, is being recorded. Unsurprisingly, the waveform does not look quite like a true transmembrane action potential recording: rather than consisting of a single positive deflection the action potential has both upward and downward components, i.e. it is not *monophasic* but *biphasic*.

If the electrode at **B** records a potential positive relative to that at **A**, this results in an upward deflection, while if the electrode at **B** records a potential negative relative to that at **A** this results in an downward deflection. Before the action potential arrives at **A**, both **A** and **B** are at the same potential and the recording is at baseline. When the action

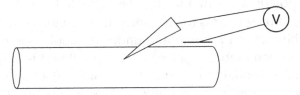

Fig. 23. Transmembrane recording of monophasic action potential.

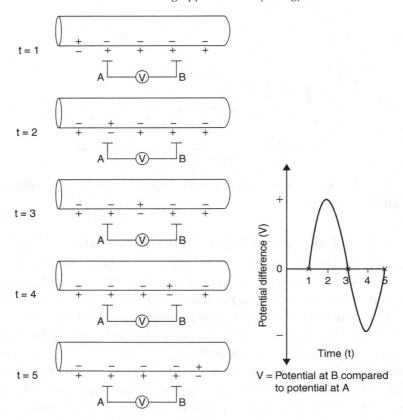

Fig. 24. Extracellular recording of biphasic action potential.

potential arrives at **A**, the region of membrane under **A** depolarises making the outside of the membrane negative. The potential at **B** is then positive relative to the potential at **A** and an upward deflection is recorded (1). The action potential then travels past **A** and the membrane under **A** repolarises again making the potentials at **A** and **B** the same: the recording returns to baseline (2). When the action potential reaches B the region of membrane under **B** depolarises making the outside of the membrane negative. The potential at **B** is then negative relative to the potential at **A** and a downward deflection is recorded (3). This gives a *biphasic* action potential recording. Incidentally, the negative component can be removed by crushing the nerve between

A and B such that the depolarising wave cannot propagate past A and the potential at B never becomes negative relative to that at A. An unexpected property of cell membranes can be exploited to obtain such recordings. Most cell membranes possess stretch-activated cation channels and therefore pressing gently on a membrane results in a local depolarisation and prevents action potential propagation past that point. This is referred to as the *monophasic action potential technique* and allows action potentials to be fairly safely recorded from the living human heart!

A mathematical treatment gives some interesting results. Let us begin by considering the current flowing longitudinally along a unit length of the axon as a result of the potential difference between 0 and x:

$$i_{long} = \frac{V_x - V_o}{r_i}$$

where r_i = intracellular resistance of the nerve axon per unit length $(\Omega \cdot m^{-1})$

i_{long} = longitudinal current per unit length $(A \cdot m^{-1})$.

If x is brought closer and closer to 0 until the distance between them is infinitely small then:

$$V_x - V_o = \frac{dV}{dx}$$

so

$$i_{long} = \frac{1}{r_i} \cdot \frac{dV}{dx}$$

Now

$$\frac{dV}{dt} = \frac{dV}{dx} \cdot \frac{dx}{dt}$$

and $\frac{dx}{dt}$ is conduction velocity, θ $(m \cdot s^{-1})$, so

$$\frac{dV}{dx} = \frac{1}{\theta} \cdot \frac{dV}{dt}$$

and

$$i_{long} = \frac{1}{r_i \theta} \cdot \frac{dV}{dt}$$

Considering the potential difference between the two extracellular electrodes, V_o, will depend on the current, i_{long} and the extracellular resistance:

$$V_o = i_{long} \cdot r_o$$

where r_o = extracellular resistance of the nerve axon per unit length ($\Omega \cdot m^{-1}$)

$$V_o = \frac{r_o}{r_i \theta} \cdot \frac{dV}{dt}$$

So V_o does not actually depend on V, but rather on the rate of change of V with time, $\frac{dV}{dt}$. Clearly, this is first positive (depolarisation) and then negative (repolarisation), explaining the biphasic nature of extracellular action potential recordings in nerves.

1.7.1 *Unmyelinated axons as cables*

A cable provides a very satisfactory model of an unmyelinated nerve axon. If a potential difference is applied transversely across the cable at its centre, then current will travel in two dimensions, transversely across the cable (representing the membrane) and longitudinally along the cable (representing the length of the axon). The longitudinal flow of current can be thought of as occurring through a resistor representing the intracellular space, with a resistance per unit area r_i. The transverse leak of current across the membrane can also be thought of as occurring through a resistor, with a resistance per unit area r_m. We must also build a capacitor into our model (Fig. 10), as cell membranes are able to store charge. A nerve axon can therefore be thought of as a network of resistors and capacitors. In this sense, the axon resembles an insulated cable and we can therefore apply *cable theory*, first developed to analyse signal transmission along undersea telephone cables.

The cable equation

Let us imagine a current is passed across the cable at 0 to set up a potential difference V_0. The current will decay as it moves along the axon as it splits across the two possible paths and we can represent the potential difference at x as V_x. Since the membrane also includes a capacitance component, V must also vary with time as the capacitance discharges. Here c_m represents the capacitance of the membrane per unit area. An analysis of this system requires us to consider how V varies with both x and t.

Let us think first of the current following transversely across the membrane, i_{trans}. This will be the sum of the current flowing through r_m and the current flowing through c_m:

$$i_{trans} = \frac{V}{r_m} + c_m \frac{\delta V}{\delta t}$$

As established in the previous section, a proportion of the total current is lost across the membrane. Therefore the longitudinal current, i_{long}, must decline as we move away from 0 such that:

$$i_{long} = \frac{1}{r_i} \cdot \frac{\delta V}{\delta x}$$

Now, if we assume that the total current remains constant then:

$$i_{trans} = \frac{\delta i_{long}}{\delta x}$$

and therefore combining these two equations:

$$i_{trans} = \frac{1}{r_i} \cdot \frac{\delta^2 V}{\delta x^2}$$

We can now put this back into the first equation to get:

$$\frac{V}{r_m} + c_m \frac{\delta V}{\delta t} = \frac{1}{r_i} \cdot \frac{\delta^2 V}{\delta x^2}$$

and this rearranges to

$$V = \frac{r_m}{r_i} \frac{\delta^2 V}{\delta x^2} - r_m c_m \frac{\delta V}{\delta t}$$

which rigorously and completely describes the effects of x and t on V in our model. Notably this model can be modified to describe other excitable tissues. For example, the standard model of a skeletal muscle fibre also includes an additional resistance and capacitance in parallel to represent the transverse tubules (see Chapter 3).

Unfortunately, this equation is not easy to solve. However, it does become much simpler if we consider the special case where a depolarisation results from a constant current applied across the cable for an infinitely long time. Many books make this assumption, despite it not really being valid. Having been made, the resulting equation can then be used to develop the attractively simple notion that the conduction velocity along a nerve fibre is directly proportional to the square root of its radius. Since this argument is so often made, we provide the underlying mathematics in the Appendix.

CHAPTER 2

Muscle as an Excitable Tissue

We saw in Chapter 1 how electrical signals are set up and passed from one excitable cell to another. In this chapter, we will consider muscle as a tissue that uses this electrical signal to trigger the conversion of chemical energy (in the form of adenosine triphosphate, ATP) into mechanical energy (work). Notice that we have not used the word *movement*. Although the primary function of muscle is to generate movement, either moving one part of the body in relation to another or moving the whole body in space, muscle can also do work without producing movement. As an example, think of the mechanical work involved in carrying a suitcase. The chemical reactions occurring in muscle also generate heat. Although this cannot then be converted to *useful* mechanical work, it is important in the maintenance of body temperature.

2.1 Basic Properties of Muscle

There are three types of muscle.

(1) **Skeletal muscle**, attached to joints via tendons, which is under voluntary control and is responsible for movement and providing support for the body.
(2) **Smooth muscle**, found within the walls of hollow organs, which is not under voluntary control and is responsible for altering the diameter of these organs and sometimes for propelling their contents.

(3) **Cardiac (heart) muscle** which is again not under voluntary control and is responsible for pumping blood to the tissues (see Chapter 5).

The arrangement of the force-producing components of these different types of muscle, as well as the detailed processes by which contraction and relaxation are regulated, vary. However, they all share the same basic machinery and contract by the same *sliding filament mechanism*.

To understand this, it is necessary to appreciate that each muscle consists of a large number of individual cells, or *fibres*. Each fibre is in turn made up of a number of parallel subunits known as *myofibrils*. These contain the contractile proteins, or *myofilaments*, which are responsible for force production, as well as regulatory proteins. The detailed composition and structure of these filaments is beyond the scope of this book. In summary, however, there are two types of myofilaments: thick and thin. Thick filaments are made up of *myosin* molecules which consist of *heavy* and *light chains*. These chains associate around a central *titin* molecule to form a long tail and globular head. Thin filaments are made up of units of *actin*, together with regulatory proteins. The two types of filament overlap each other and myosin heads are able to bind to actin and form *crossbridges*. Muscle contraction occurs when myosin and actin slide past each other, causing the muscle fibre to shorten and contract (hence the *sliding filament* mechanism). It is important to note that neither the thick nor the thin filaments themselves change length during contraction.

At the molecular level, this sliding movement arises as a result of each myosin head tilting through approximately 45° while attached to actin. This is shown schematically in Fig. 25.

Each muscle contraction requires multiple cycles of the myosin heads attaching, tilting, detaching and then reattaching. If all the myosin heads went through this cycle at the same time then a contraction would consist of a series of individual jerky shortenings. To prevent this, the myosin binding sites on actin are slightly offset with respect to the separation of myosin heads. This means that at any

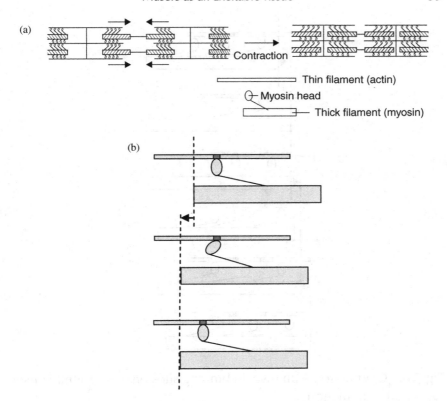

Fig. 25. The sliding filament mechanism.

given time only a proportion of myosin heads are attached and while these are detaching and reattaching, other myosin heads can bind. The individual movements are therefore averaged out and contraction smoothed. This is shown in Fig. 26.

2.2 Mechanical Properties of Muscle

These basic principles are all that is needed to understand the key mechanical properties of muscle. Starting with the fact that each crossbridge produces the same unit force (or *tension*), intuitively the total tension generated will be proportional to the number of crossbridges attached. Maximum tension will be generated at the

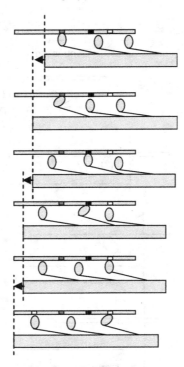

Fig. 26. Contraction with myosin binding sites on actin slightly offset to smooth contraction.

particular muscle length that results in maximum overlap between the actin and myosin filaments. As muscle length is increased beyond this optimal value, progressively fewer actin and myosin heads will overlap and tension will fall. As muscle length is decreased below this optimal value, maximum force will again fall as collisions will take place, first between the thin filaments, and then between the thick filaments. This is shown in Fig. 27.

From this, we can see that the maximal tension is actually only developed over a relatively narrow range of lengths. In skeletal muscle, this corresponds to the physiological working range. It follows that under normal conditions, the force developed within the muscle does not depend on its length. By contrast, instead of working over this plateau range, cardiac muscle normally works on the upward-sloping

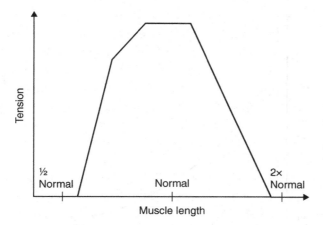

Fig. 27. The relationship between muscle length and tension in skeletal muscle.

section of the length-tension curve. This results in the force generated by cardiac muscle increasing with increasing fibre length as the walls of the heart are stretched. This forms the cellular basis of the *Starling mechanism* and helps to explain why the force of contraction of the heart depends on how much blood is returned to the heart (the *pre-load*), which in turn determines how much the fibres are stretched. We shall see in Chapter 5 that this property is fundamental to the control of the circulation as it allows control over the output of the heart on a beat-to-beat basis.

From a mechanical perspective, it is also intuitive that increasing the force against which the myosin head has to shorten will increase the average time taken for tilting to take place. This in turn will decrease the velocity of contraction until, when the force is large enough, no net movement of crossbridges will take place. At this point, contraction velocity will fall to zero. This can be seen in Fig. 28.

Although this looks rather like force and velocity might be inversely proportional, this is not in fact the case. An inversely proportional relationship would imply that the velocity of contraction does not fall to zero until the force developed is infinite: the muscle fibre would snap long before this point is reached! A. V. Hill demonstrated in 1938 that, rather than being inversely related, force and velocity are in fact

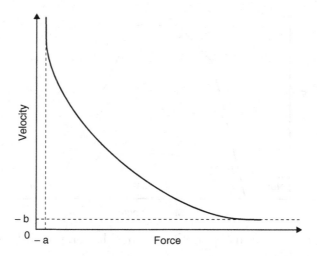

Fig. 28. The relationship between the force or load on a muscle and the velocity of contraction.

related by a hyperbolic function described by the equation:

$$(F + a) \cdot (v + b) = (F_{max} + a) \cdot b.$$

This curve has limits at $F = -a$ N and $v = -b$ ms^{-1}, i.e. $-a$ N and $-b$ ms^{-1} are the maximal forces and velocities respectively. The values of a and b for a particular muscle preparation are determined by experiment.

We can also use this relationship to understand how tension and velocity affect the work done per unit time (power output). When force and velocity are constant, the power output of a system is given by:

$$P = Fv,$$

where P = power output (W)
       F = force (N)
       v = velocity (m·s^{-1}).

From Fig. 29 we can see that, as force and velocity increase, power initially rises but then reaches a maximum and starts to fall. For any given muscle, there is therefore a particular force and velocity at which

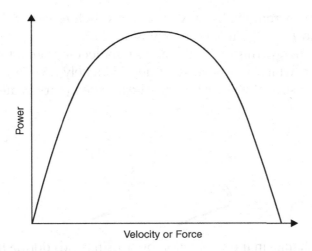

Fig. 29. The relationship between muscle power, force and velocity.

power output is maximal. An everyday example of this can be seen when riding a bicycle. Here gears enable the ratio of rotations of the pedals to rotations of the wheels to be altered as the slope of the road changes. This allows the cyclist to maintain a constant force and velocity on the pedals and therefore keep power output at a maximum.

2.3 Energetics of Muscle Contraction

Having seen that muscle contraction ultimately arises as a result of the tilting of myosin heads while they are attached to actin, we can now consider what brings about this crucial conformational change. Central to this is the concept that all systems tend towards the state of lowest free energy. A myosin molecule is nothing more than a group of atoms bonded together. If we assume that the molecule is in thermodynamic equilibrium with its environment, it will therefore tend towards a shape in which the atoms expend the least possible energy to maintain their structure. If energy is added to the system the equilibrium is altered and the myosin molecule will be driven into a higher energy state. As a consequence, any increase in the energy of the myosin molecule will tend to alter its shape. As with most processes in the body, the energy for this comes from the hydrolysis of ATP, which results in the release of ADP and P_i. ATP binds to the myosin head and is then hydrolysed to release energy. This drives the myosin molecule from a low-energy state (M) in which the head is at approximately 45° to actin, to a high-energy state (M*) in which the head is at approximately 90° to actin (Fig. 30).

$$M\text{-}ATP \longrightarrow M^*\text{-}ADP\text{-}P_i$$

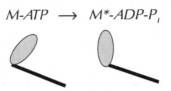

Fig. 30. Change in the orientation of myosin heads during hydrolysis of ATP.

When this high energy myosin (M*) binds to actin (A), an *actomyosin complex* is formed. This is followed by the sequential dissociation of P_i and ADP, with the conversion of myosin back to its low-energy state (M) (Fig. 31). A useful analogy is that the binding of ATP to myosin cocks the trigger, then the binding of myosin to actin to form the actomyosin complex pulls the trigger.

$$(M^*\text{-}ADP\text{-}P_i) + A \longleftrightarrow (M^*\text{-}ADP\text{-}A) + P_i$$

$$(M^*\text{-}ADP\text{-}A) \longleftrightarrow (M\text{-}A) + ADP$$

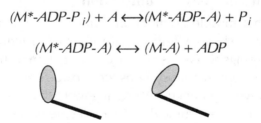

Fig. 31. Change in the orientation of myosin heads with dissociation of ADP.

The actomyosin complex then remains intact until ATP becomes available and binds to myosin, displacing actin from the myosin head.

$$(MA) + ATP \leftrightarrow (M\text{-}ATP) + A$$

It is this final step that explains why it is often said that ATP is required for muscle relaxation and underlies the development of *rigor mortis* after death (see Clinical Box 4).

Clinical Box 4: *Rigor Mortis*
After death all reactions tend towards equilibrium. Among the first of these processes is that of ion equilibration between compartments of the body as ion pumps loose their energy supplies. In the case of muscle, this results in Ca^{2+} leaking into the cell, raising $[Ca^{2+}]_i$ to high levels. The resulting uncontrolled muscle contraction hastens the total exhaustion of ATP supplies and ends with nearly all myosin molecules being in cross-linked actomyosin complexes. *Rigor mortis* describes this rigid state of muscles that develops shortly after death due to this large proportion of thick and thin filaments in cross-linked states.

2.4   Excitation-Contraction Coupling

So far we have considered how chemical energy from ATP is converted into mechanical energy during the process of muscle contraction. However, all living cells contain relatively abundant ATP, and therefore everything we have considered so far is energetically favourable. We must now ask what *prevents* these processes from occurring in resting muscles and, from this starting point, how we are then able to initiate and control contraction.

In all types of resting muscle, myosin heads are blocked from interacting with actin and therefore contraction does not occur. This block is released by Ca^{2+}. As we shall see, the nature of this blocking and the origin of the intracellular Ca^{2+} ions vary between the three muscle types. In all cases however, the final common pathway to contraction is an increase in $[Ca^{2+}]_i$. This is achieved through a process termed *excitation-contraction* (EC) *coupling*. Continuing our analogy, increasing $[Ca^{2+}]_i$ releases the safety catch on the gun.

In this section, we will first consider EC coupling in skeletal muscle and then compare this to the equivalent process in smooth and cardiac muscles. In doing so, we will highlight several of the key differences between these muscle types.

2.4.1   Excitation-contraction coupling in skeletal muscle

In skeletal muscle, contraction is prevented at rest by *tropomyosin*. This protein sits in a groove between the contractile filaments and physically blocks the myosin heads from binding to actin. In order for contraction to occur, tropomyosin must move deeper into the groove, uncovering the actin binding sites. This movement is brought about by Ca^{2+} binding to another protein, *troponin*. This causes a change in the conformation of the troponin molecule which leads to a local interaction that pulls on the tropomyosin molecule.

In skeletal muscle, this Ca^{2+} comes from intracellular stores within the *sarcoplasmic reticulum* (SR). The release of Ca^{2+} is triggered by the arrival of an action potential. Since the surface area of a fibre is rather small in relationship to its volume, it follows that a surface action potential would need to travel a fairly long distance through the cell to reach the SR. To speed up this process, inward-foldings of the skeletal muscle cell membrane, referred to as the *transverse-tubules* (T-tubules), allow action potentials on the surface membrane to spread rapidly into the depths of the fibre. As an action potential travels down the T-tubules, voltage sensors in the T-tubular membrane detect the resulting change in membrane potential. These are in fact voltage-activated Ca^{2+} channels modified to allow very little Ca^{2+} to pass through the membrane: they are referred to as *dihydropyridine receptors* (DHPRs) as they are blocked by a clinically important class of drugs, the *dihydropyridines*. In skeletal muscle, DHPRs are physically coupled to Ca^{2+} channels in the SR membrane known as *ryanodine receptors* (RyRs), so called as they are opened by the experimental drug *ryanodine*. By now, it will not come as a surprise that the change in membrane potential alters the shape of these DHPRs and this in turn results in the opening of the RyRs and in turn initiates the release of Ca^{2+} from the stores into the

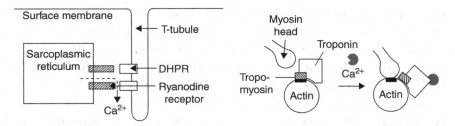

Fig. 32. Excitation-contraction coupling in skeletal muscle.

intracellular fluid. Each action potential therefore results in $[Ca^{2+}]_i$ increasing. This Ca^{2+} binds to troponin, moving tropomyosin out of the groove, thus allowing myosin and actin to interact and generate force (Fig. 32).

2.4.2 *Excitation-contraction coupling in smooth muscle*

Although the basic principles of contraction are the same in smooth muscle as in skeletal muscle, smooth muscle has a different role to play and therefore shows different adaptations. While skeletal muscle is designed to produce a force acting in a single direction at a joint, smooth muscle must often produce a circumferential force in the wall of a hollow organ. Furthermore, smooth muscle is not required to contract against any greater load than that produced by the pressure within the hollow organ. In contrast to skeletal muscle, smooth muscle is therefore adapted in such a way that it is able to contract against small forces for prolonged periods of time while expending relatively little energy.

The ability of smooth muscle to generate a circumferential force is a result of the anatomical arrangement of its fibres whilst the ability of smooth muscle to produce small forces for long periods with little energy expenditure is achieved as a result of its complex EC coupling cycle. Spontaneous contraction is still blocked by preventing the binding of actin and myosin: the protein *caldesmon* binds to actin and inhibits crossbridge cycling in the same way as troponin blocks the binding sites on actin in skeletal muscle. As with skeletal muscle, this step is also controlled by Ca^{2+}. However, unlike skeletal muscle this Ca^{2+} comes mostly from the outside of the cell and the entry of Ca^{2+} can be triggered by circulating hormones as well as nerve action potentials. Instead of binding directly, Ca^{2+} acts indirectly both by activating the regulatory protein *calmodulin* which binds caldesmon and by activating protein kinase C, which phosphorylates caldesmon. Both these effects cause caldesmon to dissociate from actin, uncovering myosin binding sites.

This is where the cellular control of skeletal muscle contraction stops: once the process has been initiated, a full force contraction will occur. In contrast, smooth muscle is able to modulate the resulting crossbridge cycling. In this case, both the force generated by, and the velocity of, crossbridge cycling are affected by $[Ca^{2+}]_i$. These parameters are affected both by Ca^{2+} directly through its binding to the myosin light chain and indirectly through activation of the phosphorylating enzyme *myosin light chain kinase* by the Ca^{2+}-calmodulin complex. The effect of such phosphorylation on force and velocity are shown in Fig. 33. Furthermore, if the myosin

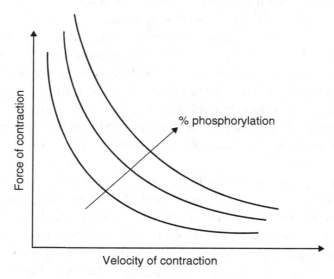

Fig. 33. The effect of phosphorylation on the force and velocity of contraction in smooth muscle.

light chain is dephosphorylated when attached to the thin filament, it remains strongly bound. Cross-bridges attached in this state are termed *latch bridges*. As a result of the prolonged time it takes for latch bridges to detach without consuming ATP, smooth muscle can remain under tension for long periods of time without much additional energy expenditure.

As might be expected, these additional biochemical pathways slow down responses in smooth as compared to skeletal muscle. However, they also allow smooth muscle function to be modulated by circulating hormones, a property absent from skeletal muscle which is exclusively under direct neuronal control. Furthermore, over the long term the levels of phosphorylating and dephosphorylating enzymes can be altered by the regulation of gene transcription. We will see in Chapter 5 that these forms of regulation are particularly important for smooth muscle in the walls of blood vessels.

2.4.3 *Excitation-contraction coupling in cardiac muscle*

EC coupling in cardiac muscle is similar to that in skeletal muscle: the interaction between actin and myosin is blocked by tropomyosin, and this block is removed by Ca^{2+} binding to troponin in the same way. In ventricular muscle cells, the increase in $[Ca^{2+}]_i$ required is again initiated by an action potential travelling down T-tubules and causing a conformational change in the DHPR. In contrast, atrial cells have a relatively large surface area to volume ratio and, probably for this reason, lack a T-tubular system. In atrial cells, the DHPR is located on the surface membrane. In such cardiac muscle, the DHPR behaves as a true voltage-activated Ca^{2+} channel: depolarisation opens the DHPR resulting in the influx of Ca^{2+}. This Ca^{2+} then acts on RyRs to cause Ca^{2+}-induced-Ca^{2+} release (CICR), an unusual example of positive feedback. Ca^{2+} influx from the extracellular fluid and Ca^{2+} release from the SR are thus both important: the former triggers the latter. Anything that results in an increased influx of Ca^{2+}, such as increased $[Ca^{2+}]_o$ or longer cardiac action potentials, will increase $[Ca^{2+}]_i$ and in turn increase the amount of Ca^{2+} in the stores able to be released with the next action potential.

Clinical Box 5: *Mechano-electrical Feedback*

Accounts of this area usually focus exclusively on the role of electrical events in triggering mechanical events. However, it is well established that the reverse also occurs. This is termed *mechano-electrical feedback* and probably results from the ion channels we have described being sensitive to stretch. While relatively little research has been conducted on such mechanisms they have at least two clinically important consequences with regard to the heart. First, it is sometimes possible to cause life-threatening cardiac arrhythmias to revert to normal rhythms by striking a patient's chest (*precordial thump*). Secondly, such abnormal rhythms sometimes result from blows to the chest (*commodio cordis*). This is a surprisingly common cause of sudden cardiac death in athletes. It also has important consequences in the study of cardiac electrophysiology. This is because several recording techniques in common use require that hearts be pretreated with electromechanical uncoupling drugs in order that contraction does not result in movement. This might otherwise produce recording artefacts.

2.5 The Control Over $[Ca^{2+}]_i$ and Force Production

At this point, it is worth taking a closer look at the origin of the Ca^{2+} determining $[Ca^{2+}]_i$ in each of the three muscle types as it has important implications for the grading of force. In skeletal muscle, $[Ca^{2+}]_i$ is determined almost exclusively by the release of Ca^{2+} from the SR triggered in response to surface action potentials. After each contraction, Ca^{2+} is then taken back up into the stores by an ATP driven Ca^{2+} pump (the *smooth endoplasmic recticulum Ca^{2+} pump*, SERCA), returning $[Ca^{2+}]_i$ back to the low resting level. Skeletal muscle therefore has a closed Ca^{2+} economy entirely within the cell: with each contraction a supramaximal burst of Ca^{2+} is released from the stores. The result of this is that little, if any, control is exerted over the amount of Ca^{2+} released. This means that the arrival of an action potential at a muscle fibre will always result in the same force being generated by that fibre.

In order to increase the force generated by a given muscle, it is necessary to increase the number of fibres contracting — a strategy referred to as *recruitment*. To understand this, it must be appreciated that each individual fibre is innervated by a single motor neuron and the group of fibres activated by a single motor neuron is called a motor unit. All fibres in a motor unit are of a similar type. Not surprisingly, the largest diameter muscle fibres that produce the largest forces are innervated by large diameter neurons and the smallest diameter muscle fibres producing smaller forces by small diameter neurons. As shown in the Appendix, if a cell membrane is treated as a parallel-plate capacitor then its capacitance is proportional to the membrane area. The higher the capacitance of the membrane, the more charge must cross to produce a unit change in the transmembrane potential difference, i.e. to result in depolarisation. It follows that smaller cells, with a smaller membrane surface areas and lower membrane capacitances, are stimulated by smaller post-synaptic currents (see Chapter 1) and so are recruited first. As the size of the post-synaptic current increases, the larger muscle fibres which produce greater force are then progressively recruited.

In contrast, both smooth and cardiac muscles have open Ca^{2+} economies. The increase in $[Ca^{2+}]_i$ initiating contraction in smooth muscle results predominantly from Ca^{2+} entry from the outside of the cell; cardiac muscle makes use of both intracellular and extracellular Ca^{2+}. Additionally, in both these cases, all the individual muscle cells are joined together and the whole muscle contracts as a whole (a *syncytium*). Recruitment of additional cells is therefore not possible and modulation of force must instead take place at the level of the single cell. As we have already seen, smooth muscle uses Ca^{2+}-dependent phosphorylation to alter the force generated by individual myosin crossbridges and cardiac muscle regulates force by the Starling mechanism, among other means. Furthermore, unlike in skeletal muscle, increases in $[Ca^{2+}]_i$ following action potentials in cardiac muscle are submaximal. Cardiac muscle is therefore able to

Clinical Box 6: *The Ryanodine Receptor and Abnormal Cardiac Rhythms*

Digoxin, a drug formerly used to treat cardiac failure, may work by increasing the leakiness of the RyR. This would serve to increase $[Ca^{2+}]_i$ and thereby increase the force produced during contraction. However, digoxin has fallen out of favour for use in treating cardiac failure as it is often associated with abnormal cardiac rhythms (*arrhythmias*), which may occasionally impair the ability of the heart to pump blood and therefore be life threatening. This is probably explained by an inbalance between mechanisms elevating $[Ca^{2+}]_i$ and mechanisms returning $[Ca^{2+}]$ to its resting level.

In most cell types, Ca^{2+} is pumped outwards across the membrane by the ATP-consuming *plasma membrane* Ca^{2+} *pump* (PMCA). This extrudes one Ca^{2+} ion in exchange for two H^+ and therefore does not result in the net movement of charge across the membrane. However, if this were the sole mechanism in place to extrude Ca^{2+} from cardiac cells, then the large increases in $[Ca^{2+}]_i$ that trigger cardiac muscle to contract would result in the PMCA consuming more ATP than the cell could possibly produce over the relevant timescale. Cardiac muscle cells therefore require a second Ca^{2+} extrusion mechanism, the Na^+/Ca^{2+} exchanger (NCX). This is only activated when $[Ca^{2+}]_i$ becomes high. The NCX extrudes one Ca^{2+} ion in exchange for the entry of three Na^+ ions: this "borrowed" energy must be "paid back" by the Na^+/K^+-ATPase (see Chapter 1) over time. Each pump cycle therefore results in the net movement of one unit charge into the cell. As discussed in Chapter 1, this results in depolarisation. Abnormal increases in $[Ca^{2+}]_i$ may thus result in abnormal depolarisations, leading to arrhythmias.

Abnormal leakiness of the RyR might also explain life-threatening abnormal rhythms in the rare hereditary disorder *catecholaminergic polymorphic ventricular tachycardia* (CPVT). In sufferers of CPVT, arrhythmia is often associated with exercise or emotional events that result in increased circulating concentrations of adrenaline and other cathecholamines. This may result in phosphorylation of RyRs in cardiac muscle cells, increasing their leakiness. Abnormal RyR phosphorylation is also associated with heart failure.

regulate force by altering the amount of Ca^{2+} in the SR, as well as the amount entering from outside the cell, during an action potential. It should be noted, however, that such changes tend to affect the force of the next, rather than the current, contraction.

CHAPTER 3

The Digestive System

Every cell in the body requires an adequate supply of energy to survive and function normally. Within individual cells this energy is almost exclusively provided in the form of high-energy covalent bonds in adenosine triphoshate (ATP) molecules. Energy is liberated from this ATP by the removal of one of the inorganic phosphate groups, generating adenosine *di*phosphate (ADP). ADP can then be reconverted back to ATP by glycolysis (the process by which glucose, a six-carbon sugar, is broken down into two three-carbon pyruvate molecules) and the Krebs cycle (the final common pathway for the processing of fuel molecules). These pathways are beyond the scope of this book but the substrates for these processes, and therefore the ultimate source of energy for the body, come from the food we eat. Leaving aside the complexities of cellular metabolism, carbohydrates are the body's first source of energy and can be used in both glycolysis and the Krebs cycle. When carbohydrate supplies run low fats are digested into glycerol, which enters glycolysis, and fatty acids, which enter the Krebs cycle. Proteins (long chains of amino acids), which are able to enter the Krebs cycle, are the final fuel to be used. Eating is not just about acquiring energy, however. Protein, fats, vitamins and other substances found in food are also of great importance.

Before food can be used by the body, it must first be broken down (*digested*) from larger, more complex molecules to smaller, simpler molecules. These smaller molecules can then be absorbed into the blood stream and ultimately transported to individual cells. When these reach their destinations, they can act either as substrates for the production of ATP or as building blocks for growth and repair. The body never absorbs a large protein from the gut and then plugs it directly into a structure. Large molecules are always broken down to small molecules and then rebuilt.

In this chapter, we consider the role of the digestive system in this process. We cover neither individual processes nor descriptive biochemistry in detail, but instead give a broad overview.

3.1   The Overall Structure of the Digestive System

At its simplest level, the digestive system consists of a series of hollow tubes running from the mouth to the anus (the *gut*). It also includes two solid organs, the liver and the pancreas (Fig. 34). These components of the gut produce various secretions, including enzymes and signalling molecules. Food enters through the mouth and is digested into smaller

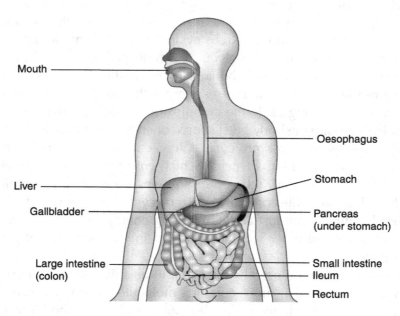

Fig. 34. The digestive tract.

molecules in a number of steps. These are absorbed and can then be either used or stored. The substances that we cannot use (or are unable to absorb) pass through the entire tube and ultimately out of the body. Unlike the kidney, which is able to vary the rates of secretion and reabsorption of substances depending on the requirements of the body at that time, there is no such regulation in the digestive system. Indeed, in the normal functioning digestive system, the useful components of food are all broken down and absorbed, regardless of the nutritional state of the body. Any excess is stored for future use or removed from the body by the kidney (see Chapter 6). This ability to store energy-rich substrates in times of plenty and then draw on these stores in times of need frees us from the need to constantly eat. However, it also means that consumption of excess carbohydrate, protein or fat results in their conversion into fatty acids, which can be stored in virtually unlimited quantities. Variations in intake therefore have profound effects on body weight and composition.

Each section of the digestive tract is specialised to a particular function, just as in the kidney. For example, the stomach primarily acts as a mixing tank and a *sphincter* (valve) between the stomach and small intestine regulates the forward-movement of dissolved food. In the first part of the small intestine (the *duodenum*), secretions from the pancreas and liver mix with food. The other two sections of the small intestine, the *ileum* and *jejunum*, are the sites of final digestion and most of the absorption of organic substances. The large intestine then absorbs ions and water with the rectum serving as a final storage site to allow intermittent emptying of the digestive tract. It is no coincidence that ions and water are left until the end. The basic strategy in the gut, not unlike that in the kidney (Chapter 6), is to digest all the food first, absorb the breakdown products that are useful and then deal with the osmotic problems at the end. It makes sense that it should be this way around because the end osmotic result of digestion and absorption is so unpredictable. The fact that water is left until the end also has the added advantage of maintaining bulk flow through the system.

At this point, the liver deserves a mention. One key role of the liver is to secrete *bile*, which as we will see below is important for the digestion of fats. Many of the components of bile are ultimately excreted from the body, providing an opportunity for toxic substances derived from the breakdown of haemoglobin to be lost. A second role of the liver is to reprocess material absorbed from the gut. Indeed, the blood supply to the liver is arranged in series with that to the gut (the *hepatic portal circulation*) and therefore everything that is absorbed within the digestive tract passes through the liver. This is important as if all products of the unregulated process that is digestion were dumped directly into the systemic circulatory system severe consequences for the maintenance of homeostasis could result. The liver also detoxifies potentially toxic substances, including alcohol and other drugs. This is of great advantage if the substance is harmful but may be less helpful in the case of therapeutic drugs. Indeed, some drugs are completely inactivated in one pass through the liver and so cannot be given by mouth. The liver also serves a reservoir function, storing excess carbohydrate as glycogen in times of plenty and then releasing it in times of need.

3.2   Gut Motility

It is important that the contents of the gut are propelled at appropriate speeds to facilitate the complex processes of digestion and absorption. In fact, gut motility is so important that the digestive system is lined with smooth muscle and specialised cells and has its own nervous system, the *enteric nervous system*. This consists of more than one hundred million neurons and is often referred to as the *second brain*. *Afferent* (input) neurons link directly to *efferent* (output) neurons via interneurons. These efferent neurons innervate smooth muscle in the walls of the gut, as well as the cells that secrete enzymes and signalling molecules.

The contents of the gut move as a result of periodic, coordinated contraction and relaxation of smooth muscle in the gut wall. The key to the control of propulsion are the *interstitial cells of Cajal*. These behave as pacemakers, just like the cells in the sinoatrial node in the heart, generating oscillations in resting membrane potential (*slow*

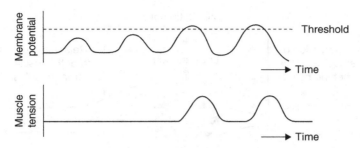

Fig. 35. The initiation of muscle contraction in the gut.

waves). If and when these slow waves reach threshold (Chapter 1), contraction is initiated (Fig. 35). Slow waves arise without any extrinsic input. However their amplitudes, and therefore the likelihood that membrane potential will reach threshold, can be modulated by nervous and chemical signals. Disruption of these electrical events may result in contraction becoming uncoordinated. It has been suggested that the unpleasant clinical syndrome of *irritable bowel disease* may result from a form of arrhythmia in gut smooth muscle (see Chapters 1 and 2).

Gut smooth muscle fibres are arranged in two specific orientations to serve particular functions. On the one hand, contraction of *circular* smooth muscle allows one part of the gut to be temporarily divided from another (as with sphincters). On the other hand, sequential contraction of *longitudinal* smooth muscle creates a pressure gradient to drive the movement of foot along the gut (compare this with muscle pumping in the venous circulation, see Chapter 5). One particularly important intrinsic reflex modulated via the enteric nervous system is *peristalsis*. This is a distinctive pattern of smooth muscle contraction stimulated by the presence of a mass of food (*bolus*) inside the gut. Local stretch of the wall of the intestine activates a reflex which ultimately results in contraction of the smooth muscle behind, and relaxation of the smooth muscle in front of, the bolus (Fig. 36).

As well as these local reflexes, longer-range reflexes allow changes in the environment in one section of the gut to influence the rate of

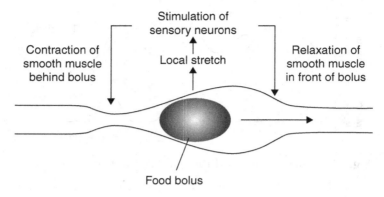

Fig. 36. Peristalsis.

movement and secretion in another. Central to this are two types of control system: *feedforward* and *feedback* (see Introduction). Regions of the gut send signals forward to sections of the gut in front of them to warn them of the imminent arrival of food and back to sections of the gut behind them to control the rate at which food reaches them. Perhaps the best example of feedback is the *enterogastric reflex*. Via this reflex various stimuli such as decreased pH, distension, breakdown products of protein digestion and changes in osmolarity, decrease the rate of gastric emptying. This allows the rate of emptying of the stomach to be controlled to allow for variations in the compositions of meals. The *gastrocolic* and *duodenocolic* reflexes provide examples of feedforward. Distension of the stomach (*gastro-*) or duodenum (*duodeno-*) with food stimulates a coordinated modified type of peristalsis from the middle of the colon (*-colic*) to the junction with the rectum known as a *mass movement*. This results in large movements of the contents of the colon — making space for the food soon to arrive from higher up the digestive tract. With the exceptions of putting food in (*swallowing*) and getting waste out (*defecation*), the gut is therefore able to propel its contents without conscious input from the brain. Higher brain centres are able to influence gut motility, though their effects are limited.

Clinical Box 7: *Diarrhoea and Constipation*

Diarrhoea describes abnormally frequent defecation with faeces of an abnormally high water content. It arises either as a result of increased secretion of water into the gut (*secretory diarrhoea*) or decreased absorption of water from the gut (*malabsorptive diarrhoea*). Secretory diarrhoea most often results from bacterial toxins, such as cholera and shigella toxins, which increase the secretion of Na^+ and Cl^- into the gut. Water then follows by osmosis. Malabsorptive diarrhoea may arise for a number of reasons. The presence of non-absorbable materials, such as the *laxative* lactulose or the sugar lactose in individuals unable to digest it, may again result in an osmotic drive for water to enter the gut. Alternatively, absorption of the circa 10 L of water that passes through the gut each day may become impaired. This may occur as a result of a decrease in the surface area of the gut wall, such as occurs in *coeliac disease*, following surgical gut resection, or when transport mechanisms are impaired by drugs. Finally, inappropriately high flow through the gut, as might occur in irritable bowel disease (see earlier), may not allow time for sufficient absorption. Diarrhoeal symptoms are treated with drugs such as loperamide that decrease gut motility. When large volumes of fluid are lost through diarrhoea it becomes essential to replace these losses with isotonic, ion-containing solutions designed for rehydration.

Constipation describes the converse abnormality to diarrhoea: defaecation takes place abnormally infrequently and faeces are hard, with an abnormally low water content. Constipation may either result from abnormalities of the composition of the faeces or the movement of faeces through the gut. Faeces often become hard as a result of an inadequate intake of dietary fibre or water, or consumption of antacids containing aluminium salts. Alternatively, decreased gut motility, perhaps occurring as a result of

(Continued)

treatment with opioid drugs (of which loperamide is an example), or anatomical abnormalities of the gut such as narrowings (*stenoses*), may impair the movement of faeces. A number of drug classes are available for the treatment of constipation. These are together referred to as *laxatives. Bulking agents* act by aiding the transit of faeces through the gut and increasing the volume of water it retains. *Osmotic laxatives* draw water into the gut and *stimulant laxatives* stimulate peristalsis.

3.3   Principles of Digestion

Having seen how food enters and moves along the digestive tract, we are now ready to consider the parallel process of digestion. As discussed above, the aim of digestion is to breakdown the large complex molecules into components that are small enough to be absorbed and transported around the body. The first stage in this process is mechanical disruption of food through chewing (*mastication*) in the mouth. This breaking down of foods into smaller pieces has two main effects. First, it allows foods that would otherwise be too big to fit through the oesophagus to be swallowed. Secondly, by breaking a large piece of food into several smaller pieces, its surface area is increased. This exposes a greater proportion of the food to digestive secretions. The remainder of the digestive process is largely chemical. Enzymes and other substances are secreted into the digestive tract to chemically alter foods by cleaving individual chemical bonds. Given the structural organisation of the digestive tract, it is not surprising that the enzymes are secreted at specific locations. Those that are responsible for the break-down of larger molecules are released first, followed later by those capable of breaking down smaller molecules. In most cases, those acting on large or medium sized molecules are secreted into the lumen of the gut to mix with its contents. In contrast, those enzymes responsible for the final breakdown steps are restricted to cell membranes or to the cells lining the digestive tract. In this way the final products of digestion are generated as close as possible to the sites at which they are to be absorbed.

Many of these enzymes are secreted as inactive precursors, known as *zymogens*. Following secretion, these enzymes require activation by either a pH change, in the case of pepsinogen in the acidic stomach, or enzymatic degradation, in the case of pancreatic secretions in the small intestine. The need for this is understandable when we remember that the enzymes are capable of breaking down the cells in the body, as well as the food we eat. Pancreatitis (see Clinical Box 8) provides an example of the consequence of releasing enzymes that are already active.

Clinical Box 8: *Acute pancreatitis*

Acute pancreatitis describes inflammation of the pancreas occurring over a rapid time frame. This may occur for a wide range of reasons, including as a result of a species of scorpion indigenous only to Trinidad! More commonly, however, acute pancreatitis results either from excessive alcohol consumption or from blockage of the pancreatic ducts by *gallstones* (usually comprised of a mixture of cholesterol and the salts of bile acids). Injury to the pancreas results in activation of the zymogen tripsinogen to form active tripsin. Not only does this begin to digest the pancreatic tissue but it activates a number of other zymogens. Activation of lipase results in the digestion of fats within the pancreas, worsening tissue damage. Activation of elastase begins to digest the walls of blood vessels, potentially leading to severe bleeding. To make matters worse, activation of kininogen to form bradykinin activates the blood clotting cascade causing the accumulation of solid masses of blood within the circulation (*thrombi*) that impair flow. These processes also result in the release of amylase into the circulation: $[amylase]_{plasma}$ constitutes a useful diagnostic marker. Acute pancreatitis is a surgical emergency and can be rapidly lethal even with treatment.

The mention of gastric pH raises a further issue: the stomach produces a number of important secretions. Surprisingly, only one of these is actually *essential* to life: *intrinsic factor*. This binds to vitamin B_{12} and allows it to be absorbed in the distal part of the small intestine, the ileum. This vitamin is an essential *co-factor* for the functions of many enzymes. Other gastric secretions include acid, pepsinogen, mucus and HCO_3^-. Acid secretion is vital for the maintenance of the low gastric pH (approximately 2–3). It also has a role in the disinfection of food before it enters the small intestine. Few organisms are capable of living in the hostile environment of the stomach. In evolutionary terms, this function becomes more important as the size of the animal increases. Small animals worry about not getting eaten while big animals, including humans, worry more about parasites. Pepsinogen is a zymogen which is activated by acid to pepsin; this breaks down proteins under acidic conditions. Finally, mucus and HCO_3^- protect the lining of the stomach from acid and pepsin. The importance of this is demonstrated by the consequences of a breakdown in this barrier (Clinical Box 9).

3.3.1 Enzymatic digestion

To understand how the various enzymes break food down, we must appreciate that food contains both water-soluble and fat-soluble components. These must be handled differently. Water-soluble foodstuffs are relatively easy to handle: they are progressively broken down by water-soluble enzymes into smaller and smaller molecules. In general, this involves first splitting molecules in the middle and then chewing away at the ends. The confusion often arises from the large numbers of different enzymes involved in this process, as each enzyme is specific to a particular carbohydrate or peptide bond. Examples of these (given simply to illustrate the point) include: trypsin which attacks peptide bonds with basic carboxyl terminals; chymotrypsin which attacks peptide bonds with aromatic carboxyl terminals; elastase which attacks peptide bonds with neutral aliphatic amino acids at their carboxyl terminals; lactase which breaks down lactose; maltase which breaks down maltose and sucrase which breaks down sucrose.

Clinical Box 9: *Stomach Ulceration*

While it is important that the stomach produces acid to facilitate digestion, it is also important that this acid is not allowed to come into contact with the gastric lining. Should the protective barrier consisting of mucus and neutralising HCO_3^- become breached, acid may erode this lining. A breach of the surface layer (*mucosa*) is referred to as an *erosion*. If this then extends deeper to breach the underlying *muscularis mucosa*, it is referred to as an *ulcer*. This not only results in severe burning pain but may also lead to severe bleeding and other potentially life-threatening consequences. Gastric ulcers are often associated with infection with an otherwise unusual species of bacteria, *Helicobacter pylori*. These bacteria produce the enzyme urease which breaks down urea into NH_3. This protects them from the otherwise hostile acidic environment in the gut and allows them to burrow into the mucosa (it also provides the basis for a breath-test used to detect the presence of these bacteria). This burrowing breaches the protective barrier covering the gastric lining, allowing acid to come into contact with the mucosa. Thrombosis and inflammation probably also contribute to ulcer formation in patients infected with *H. pylori*. Gastric ulcers resulting from such infections are treated with a combination of antibiotics and a proton pump inhibitor to decrease acid production.

The same basic principle applies to fat-soluble substances. However, as all the digestive enzymes are water-soluble, these additionally need a *detergent* to make them able to mix with (become *misible* in) water. This role is served by *bile acids*. These are synthesized from cholesterol in the liver, stored in the gall bladder and then secreted into the small intestine to be mixed with food leaving the stomach. They are *amphipathic* molecules, having both water-soluble (*hydrophilic*) and lipid-soluble (*hydrophobic*) regions. The hydrophobic regions can interpose themselves between the molecules on the surface of a fat globule while hydrophilic regions can face outwards into the water (Fig. 37). As we will see again in Chapter 4 when we consider surfactant

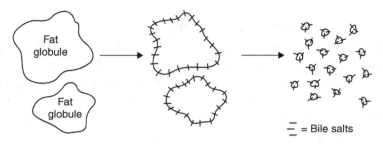

Fig. 37. The emulsification of fats by bile salts.

(another detergent), this interposition disrupts interactions between surface molecules. This reduces the force required to overcome these interactions (*surface tension*) and so allows large fat globules to be split up (*emulsified*) into microscopic droplets. This is important because it greatly increases the surface area of fat, making it available for digestion by water-soluble enzymes that cannot access the insides of droplets. Bile acids are also able to make fats soluble by forming cylinders of bile acids containing products of fat digestion (*micelles*). It is in these micelles that the products of fat digestion are transported to intestinal cells where they can be absorbed into the circulation.

3.4   Intercellular Signalling in the Digestive System

As well as this vast array of enzymes involved in the digestion of food, the digestive system also produces a range of chemical signalling molecules. These include both endocrine signals (substances secreted by one tissue that circulate in the blood stream to reach and act on a target tissue, *hormones*) and paracrine signals (substances secreted by and acting on the same tissue). These not only regulate the release of enzymes and other secretions, but also a wide range of other processes.

Most endocrine and paracrine signalling molecules in the gut are *peptides*, short sequences of amino acids. Interestingly, molecules which have very different actions often share remarkable structural similarities. The classic example is provided by the hormones *gastrin* and *cholecystokinin* (CCK). These share a *minimally active fragment* (glycine-tyrosine-methionine-aspartate-phenylalanine) that is enough alone to stimulate receptors for both hormones. Indeed, gastrin is able to act on CCK-receptors, while CCK is able to act on gastrin-receptors. This makes the number of messages that could potentially be conveyed all the larger.

Not only can a single receptor respond to multiple signalling molecules, but a single cell can express multiple receptors. The binding of signalling molecules to such receptors activates intracellular signalling cascades which often interact. Perhaps the best-studied example of this is provided by the control of gastric acid secretion (see Clinical Box 9). Acid is secreted in the stomach by parietal cells which express ATP-fuelled proton pumps on their membranes. A number of factors stimulate secretion: those of interest here are *gastrin, histamine* and ACh. Each of these signalling molecules initiates secretion when applied alone. When ACh and gastrin are applied together, the magnitude of the secretory response is additive. However, when ACh and histamine or gastrin and histamine are applied together, the resulting secretory response is far greater than the response to either of the molecules applied individually. This is seen because while ACh and gastrin initiate secretion by activating the same intracellular signalling pathway (ultimately increasing $[Ca^{2+}]_i$), histamine acts via a different pathway (ultimately increasing the concentration of another key intracellular signalling factor, *cyclic adenosine monophosphate*). Interactions between these signalling pathways are needed in order to generate a maximal secretory response: this effect is referred to as *potentiation* (Fig. 38).

This example also illustrates the importance of interactions between classes of signal. While gastrin is a hormone, histamine is a paracrine signal and ACh is a neurotransmitter. In fact, the local (enteric) and systemic (parasympathetic) nerves that secrete ACh also released a peptide signal which controls the production of gastrin (*gastrin-releasing peptide*). Such *co-transmission* is the rule in the nervous system, as alluded to in Chapter 1. There are a myriad of interactions between nervous and chemical signalling pathways, of which we have considered only a few. Interactions between mechanisms controlling gastric acid secretion are illustrated in Fig. 39.

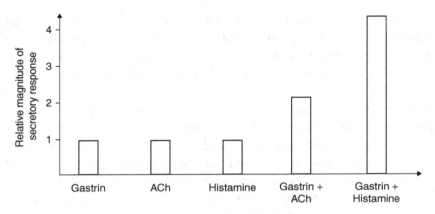

Fig. 38. Potentiation within the gut.

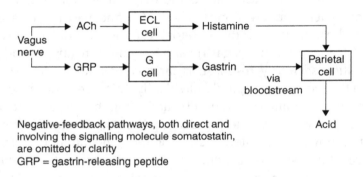

Negative-feedback pathways, both direct and
involving the signalling molecule somatostatin,
are omitted for clarity
GRP = gastrin-releasing peptide

Fig. 39. The control of gastric acid secretion.

It is often possible to deduce the functions of signalling molecules
by thinking from first principles: CCK makes a good example. CCK is
released from cells in the duodenum in response to long-chain fatty
acids. It might therefore be expected to slow gastric emptying to
allow more time for fats to be digested and absorbed, to stimulate
the pancreas to produce the enzymes required for these processes,
or to stimulate contraction of the gallbladder to release the bile
acids required for the absorption of fats. In fact, CCK does all of
these things. In doing so, it demonstrates perhaps one of the most
useful thoughts relevant to the digestive system — if it seems that

a particular molecule might do something, it probably does! Less obviously, CCK exerts a negative-feedback effect on the primordial brain centre that controls feeding (the *hypothalamus*). This prevents the digestive system from becoming overloaded. The endocrine control of food intake is discussed in Further Thoughts.

FURTHER THOUGHTS

3.5 Endocrine Control of Food Intake

It is generally accepted that homeostasis is mainly maintained by regulating output via the kidney against variable input via the gut. However, it has become clear that food intake is in fact highly regulated and such control takes place over both the short- and long-terms. A complex web of interacting neural, paracrine and endocrine signals exist for this purpose. Of these, endocrine signals are best-studied.

On the time scale of minutes to hours, the stomach produces the meal-initiator hormone *ghrelin* which act on a primal part of the brain, the *hypothalamus*, to stimulate appetite (referred to as an *orixigenic* effect). A variety of hormones, including CCK, exert a negative-feedback effect on the desire to eat (*anorectic* effect), eventually stimulating a feeling of fullness (*satiety*) and ending the meal.

Over a somewhat longer time scale, *leptin* (from the Greek *leptos*, meaning *thin*) exerts a potent anorectic effect. Leptin is produced by fat cells (*adipocytes*), with the rate of secretion varying with total body fat mass. A great deal of attention has focused on leptin in recent years as a result of the discovery of severely obese children congenitally deficient in this hormone. Some of these patients have been successfully treated with synthetic leptin and are now of a normal body mass. Interestingly, the leptin molecule is similar in structure to an important class of immune signalling molecules (*cytokines*). Indeed, had it been discovered by an immunologist it would probably be classed in this group. Leptin-deficient patients also often have immune cell defects which render them susceptible to a range of infections. Furthermore, they frequently suffer from abnormal sexual maturation, further illustrating the diverse signalling roles of leptin.

With the ever-worsening obesity epidemic, consideration has been given to the idea of administering synthetic leptin to overweight people in the general population. However, this has proved unsuccessful. Obese individuals already have high $[leptin]_{plasma}$ yet still continue to eat. This is explained by leptin resistance, similar in many ways to the insulin resistance that underlies type II diabetes. While this story is fascinating, it is beyond the scope of this book.

It is interesting to note that, while a number of anorectic hormones have been identified, comparatively few orexigenic hormones are known to exist. Furthermore, while a number of congenital over-eating (*hyperphagic*) disorders are known, no congenital under-eating (*hypophagic*) disorders have been identified. This may reflect a basal drive towards eating and weight gain, with hormones and other signalling molecules exerting a braking effect. While this drive would certainly have been useful for the survival of our prehistoric ancestors, it may also be in part responsible for the current obesity epidemic.

CHAPTER 4

The Respiratory System

All living cells have an absolute requirement for a regular supply of energy. Energy is usually handled in the form of the molecule adenine triphosphate (ATP), a product of a series of metabolic reactions within the mitochondria of each cell together referred to as *respiration*. A fuel such as glucose is the only essential substrate for respiration. However, the efficiency of the process is increased by up to eight times in the presence of O_2, which ultimately takes the role of an oxidising agent. The provision of O_2 to the mitochondria, and the removal of the waste product of respiration, CO_2, are therefore essential for life. This chapter considers how these two goals are achieved at the level of the whole organism.

The process of delivering O_2 to and removing CO_2 from the body may be split into three: transporting the gases into and out of the organism, getting the substance to or from the active cells within the organism and finally getting the substance to or from the mitochondria within each cell. In organisms such as bacteria where surface area to volume ratio is large and distances within the organism are small, simple diffusion is sufficient to solve these problems. This same strategy is used to supply O_2 to the human cornea. However, in larger multicellular organisms such as humans, distances are ordinarily limiting. We therefore require a dedicated organ, the lungs, to provide a large surface area over which gas exchange can occur. Furthermore, we also require a specialised circulatory system, the bloodsteam, to carry these gases around the body. We will begin by discussing the transport of gases into and out of the lungs, then focus on the exchange of gases between the lungs and the blood before finally considering how these gases are carried in the blood. The transport of this blood to the tissues is dealt with in Chapter 5.

4.1 The Transport of Gases into and out of the Lungs

All fluids, whether liquids or gases, will flow from regions of high pressure to regions of low pressure (see Appendix). It follows that in order for air to enter the lungs when we breathe in (*inspiration*), the pressure in the lungs must fall below atmospheric pressure. Conversely, in order for waste gases to leave when we breathe out (*expiration*), the pressure must rise above atmospheric pressure. This is achieved by using a property of gases which is described by Boyle's law. This states that "the pressure exerted by a fixed amount of an ideal gas at a constant temperature is inversely proportional to its volume". Although the gas mixture in the lungs is not ideal, most importantly because each molecule of gas has a finite volume, the principle is still approximately true and means that an increase in lung volume leads to a predictable decrease in pressure within the air spaces. Conversely, a decrease in lung volume leads to an increase in pressure. The consequence of this is that in order to generate the pressure changes needed to force air to flow into and out of the lungs, a change in lung volume is all that is needed.

Before we can begin to understand how these changes in volume are brought about, we first need to consider what determines lung volume at rest, i.e. at the end of expiration.

4.1.1 The respiratory system at rest

As Fig. 40 shows, the lungs are in a sealed chamber, the thorax. Both the lungs and the chest wall are covered by thin membranes, the *parietal* and *visceral pleuras* respectively. The narrow space between the pleuras, the *intrapleural space*, is filled with a thin layer of fluid. Strong cohesive forces generated by attractive intermolecular forces in this fluid pull the two layers of pleura together. Thus the pleural fluid ensures that if a force acts on the chest wall then it also acts on the lungs: if the volume of the thorax increases, so does the volume of the

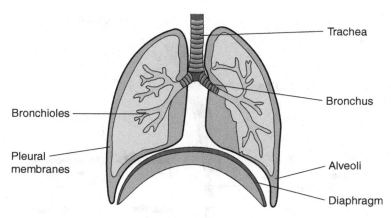

Fig. 40. The respiratory system.

lungs, and vice versa (try pulling two wet microscope slides apart). It is fair to wonder why it is necessary to have an intrapleural space filled with fluid and not instead simply have the lungs fixed to the chest wall. This might seem a particularly good alternative when we later consider the potentially devastating consequences of air entering the intrapleural space (*pneumothorax*). However, the fluid allows the pleural membranes to slide over each other as lung volume changes, greatly reducing friction forces and therefore resistance.

The lungs and the chest wall form a closed system which is in equilibrium at rest. The total inward forces must therefore equal the total outward forces, i.e. the net force acting is zero (Fig. 41). Two forces act on the lungs: a force pulling inwards due to the elastic tension in the lung tissue, which at equilibrium is stretched, and another force pulling inwards due to the surface tension in the alveoli, which we will discuss in more detail later. Both these forces tend to make the lungs collapse. These inward forces are balanced by an outward force due to the elastic tension in the chest wall, which at equilibrium is compressed. The end result is that the lungs and chest wall are pulled away from each other. This increases the volume of the intrapleural space and results in the intrapleural pressure being negative (Fig. 42).

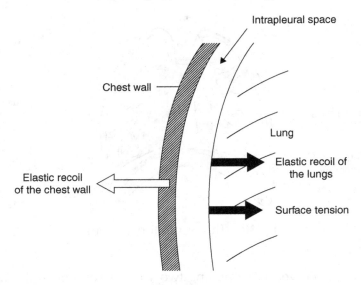

Fig. 41. Forces acting on the lungs and chest wall.

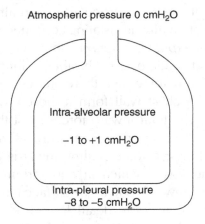

Fig. 42. Pressures in the respiratory system.

4.1.2 *Pressure changes during breathing*

So far we have shown that it is the balance of inward and outward forces acting on the lungs and chest wall which define their volumes at rest. The same forces are responsible for changes in volume: increasing

Clinical Box 10: *Pneumothorax*

If the intrapleural space becomes connected with either the atmosphere outside or with the alveoli, air will tend to enter down its pressure gradient. This can occur as a result of trauma or disease and is known as a *pneumothorax*. When this occurs, the intrapleural pressure becomes positive and, since there is no longer a force to counteract the elastic tendency of the lung to shrink, it collapses. The lung will remain collapsed until the negative intrapleural pressure is restored. If the defect is temporary, most small pneumothoracis will eventually resolve on their own as the air within the intrapleural space is gradually absorbed by capillaries surrounding the lung. However, in a sick patient with significant air in this space, a *chest drain* may be needed. In essence, one end of a tube is introduced into the intrapleural space and the other is placed under the surface of water. The water acts a valve, allowing air to escape from the intrapleural space on expiration when the pressure in the thorax exceeds atmospheric pressure, but preventing it from entering during inspiration.

the volume of the thorax increases the volume of the intrapleural space, making the intrapleural pressure more negative, increasing the outward force on the lungs and increasing lung volume.

This increase in the volume of the thorax is achieved by contracting the muscular diaphragm, which forms the base of the chest cavity, and the external intercostal muscles between the ribs, which form the walls of the cavity. The result is that during quiet inspiration the volume of the intrapleural space increases and the intrapleural pressure decreases from $-5\,cmH_2O$ at end-expiration to $-8\,cmH_2O$ at end-inspiration. This in turn leads to an increase in lung volume and a decrease in the intra-alveolar pressure from $0\,cmH_2O$ to $-1\,cmH_2O$. Air then flows down its energy gradient into the lungs until intra-alveolar pressure reaches zero at end-inspiration and air flow stops.

Conversely, expiration requires an increase in alveolar pressure above atmospheric pressure. During quiet expiration, this occurs passively as the lungs and chest wall recoil: the intra-alveolar pressure reaches a maximum of around +1 cmH_2O before again falling to zero. Thus during quiet inspiration and expiration, the pressure in the alveoli only varies between -1 cmH_2O and $+1$ cmH_2O. These very small pressure changes are all that is needed to produce the necessary changes in lung volume because the pressures are applied, and therefore the forces act, over the entire surface of the lung. This is in sharp contrast to the situation when patients are mechanically ventilated through tubes inserted into their upper airways. Then, pressure changes can only be applied at the end of the tube and these must be conducted along the airways to the lungs. The end result is that far larger pressures are needed for mechanical ventilation than normal ventilation.

4.1.3   Compliance

We can now begin to consider the importance of the physical properties of the respiratory system. We will begin with the lungs and chest wall. The stretchiness of the lungs and chest wall are usually quantified as *compliance*, the change in volume per unit change in pressure. This is represented by the gradients of the *static volume pressure curves* show in Fig. 43. To construct such static curves, data points are constructed by stopping breathing at a range of lung volumes and measuring the difference between the pressure at the mouth (atmospheric pressure) and the pressure in the intrapleural space (approximated as the intra-oesophogeal pressure). Notably, the compliance of the lungs and chest wall added together is less than the compliance of either individually. As the curves show, compliance is not constant but rather changes with volume. This is because the stretchiness of both the lungs and chest wall changes as they are stretched, just as an elastic band becomes less stretchy the more it is pulled.

As well as this stretchiness of the lung tissue, *surface tension* makes a key contribution to compliance. All molecules in a liquid are subjected to strong intermolecular attractive forces acting in all directions. Since

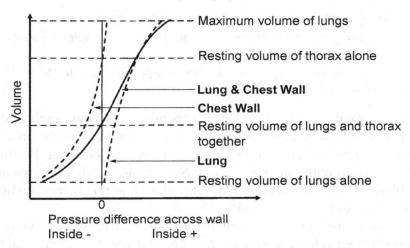

Fig. 43. Static volume pressure relationships for the lungs and chest wall.

intermolecular interactions in a gas are far weaker than those in a liquid, when a molecule is at a liquid-gas boundary attractive forces acting from below (within the liquid) are much stronger than those acting from above (with the gas). This creates a net force acting perpendicular to the liquid-gas boundary which is called the surface tension. If the boundary is curved, as is the case with alveoli, this surface tension acts towards the centre of the curvature and results in a collapsing force. If an alveolus is to be inflated, these forces must be overcome.

Classically, LaPlace's law has been applied to give a relationship between the radius of an alveolus, the surface tension and the transmural pressure difference needed to counteract this surface tension and prevent the alveolus from collapsing. LaPlace's law is effectively a statement of Newton's Third Law, "for every action there is an equal and opposite reaction". In this case two forces act, an outward force due to the transmural pressure difference across the alveolar wall and an inward force due to the transmural tension in the wall. Both of these forces are expressed per unit length. For a thin-walled sphere:

$$P_i - P_o = \frac{2T}{r}$$

where P_i = pressure inside the alveolus (Pa **or** mmHg)
P_o = pressure outside the alveolus in the intrapleural space
(Pa **or** mmHg)
T = transmural tension in the wall per unit length ($N \cdot m^{-1}$)
r = radius (m).

This is dealt with further in the Appendix. However, application of LaPlace's law requires the implicit assumptions that alveoli are perfectly spherical and are not continuous with each other. Neither of these assumptions are strictly true. Nonetheless, with these caveats in mind, we can still gain useful insights into the importance of surface tension forces by applying LaPlace's law.

Hence for a given surface tension, the transmural pressure difference $(P_i - P_o)$ needed to keep an alveolus open is inversely proportional to its radius. It follows that a small alveolus needs a larger transmural pressure difference to keep it open than does a large alveolus. Water has a surface tension of $0.07\ N \cdot m^{-1}$. Therefore for a small alveolus with a radius of $50\ \mu m$ filled with water, the transmural pressure gradient, $P_i - P_o$, needed to prevent collapse is given by:

$$P_i - P_o = \frac{2 \cdot 0.07}{50 \cdot 10^{-6}} = 2.8 \cdot 10^3\ \text{Pa}$$

and since $1\ cmH_2O \simeq 100\ Pa$

$$P_i - P_o = 28\ cmH_2O$$

At end-expiration, the intra-alveolar pressure is zero. Therefore, to prevent this alveolus from collapsing, the intrapleural pressure, would need to be huge, $-28\ cmH_2O$. In fact the intrapleural pressure is around $-5\ cmH_2O$ at end-expiration, so *something* must be lowering the surface tension. That something is *surfactant*, a complex mixture of proteins produced by a specialised class of alveolar cells which coats the inner surface of alveoli. Surfactant lowers surface tension to around $0.012\ N \cdot m^{-1}$. The most important component of surfactant, dipalmitoyl phosphatidyl choline, interposes itself between water molecules in the film covering the inner-alveolar membrane. This disrupts hydrogen bonding and reduces the forces needed to overcome the intermolecular interactions between water molecules. As a result, the intrapleural pressures needed to keep alveoli open are reduced by a factor of:

$$\frac{0.07}{0.012} = 5.8$$

meaning that the alveolus used in the example will be kept open by intrapleural pressures of $-4.8\,cmH_2O$. This is a far more realistic value. Another way of putting this is to say that surfactant greatly increases the change in lung volume produced by a given change in intrapleural pressure, i.e. it significantly *increases* lung compliance.

Changes in the arrangement of surfactant molecules with changing volume alters lung compliance. This largely explains the phenomenon of *hysteresis* (the relationship between pressure and volume seen on a static pressure volume curve being different on inspiration and expiration), much beloved of respiratory physiologists. This is certainly a valid phenomenon but only has a noticeable effect at very low lung volumes, well below the normal working range.

4.1.4 Resistance

So far we have ignored the passage of air between the atmosphere (at the mouth) and the alveoli. In fact, the airways provide considerable

---

Clinical Box 11: *Infant Respiratory Distress Syndrome*

One of many problems faced by premature neonates is *Infant Respiratory Distress Syndrome* (IRDS), a disorder resulting from the absence of surfactant leading to decreased lung compliance (this defines a *restrictive lung disease*, see later). IRDS is seen in neonates born before the 32nd week of gestation, the point at which fetuses begin to produce surfactant. IRDS can be treated by giving synthetic surfactant to the neonate through a tube inserted into the airways. Endogenous surfactant production is usually triggered by the corticosteroid surge occurring before birth. If time allows, synthetic corticosteroids can be given to the mother before delivery to stimulate surfactant production.

resistance to the flow of air. As for the flow of any fluid along a tube, the resistance along the tube is given by Poiseuilles' law (assuming that certain key conditions are met, see Further Thoughts) which tells us that:

$$R_{fluid} \propto \frac{1}{r^4}$$

where R_{fluid} = resistance to fluid flow ($kg \cdot m^{-2} \cdot s^{-2}$ **or** $mmHg \cdot L^{-1} min$)
r = radius of pipe (m)

By Darcy's law:

$$\Delta P = \dot{V} \cdot R_{fluid}$$

where ΔP = pressure difference driving flow (Pa **or** mmHg)
$\dot{Q}$ = flow along the tube ($m^3 \cdot s^{-1}$ **or** $L \cdot min^{-1}$),

so if R_{fluid} increases, the pressure gradient required to produce the same flow increases.

We can use these equations to explain why it is not possible to breathe through a long snorkel and largely why people with asthma have problems breathing. Increasing the length of the resistor through which air must flow, e.g. by breathing through a snorkel, will proportionately increase the resistance and in turn increase the overall pressure gradient needed to produce flow. The maximum intra-alveolar pressure which can be generated even on forced expiration is around $20 \, cmH_2O$ and this imposes an absolute limit on the length of a snorkel through which one can breathe. Similarly, decreasing the radius of the resistor will hugely increase in the overall pressure gradient needed to maintain flow. Asthma, the archetypal example of an *obstructive lung disease*, is characterised by airway narrowing, i.e. decreased airway radius. This can be experienced by breathing through a straw.

Since the airways are tethered to the lung tissue, as lung volume increases during inspiration r increases and, therefore, R_{fluid} decreases, as lung volume increases during inspiration. If we further assume that intra-alveolar pressure is constant during expiration then, it follows from Darcy's law that the highest air flow will be achieved when lung volume is at its highest. Furthermore, air flow will decline during expiration. A consequence of this is that healthy patients exhale the majority of the air in their lungs in the first second of a forced expiration. A decrease in this proportion is an important marker of the severity of obstructive lung diseases such as asthma which result from increases in R_{fluid}.

4.1.5 The energy needed to breathe

While the ultimate purpose of breathing is to provide O_2 and remove CO_2 and allow the continued conversion of energy into a useable form, it is important to remember that the act of breathing itself consumes energy. As shown on the next page, the total work done (or energy converted) during a respiratory cycle is given by the area enclosed within a *dynamic volume pressure loop* obtained during normal breathing (Fig. 44). Notably, the gradient of the line AB on such a curve gives the mean compliance over the respiratory cycle, while the length of the line *r* gives an index of the airway resistance. This includes both work done in stretching the lungs and chest wall, which depends entirely on compliance, and work done in moving air through the airways, which depends on resistance. We see this in

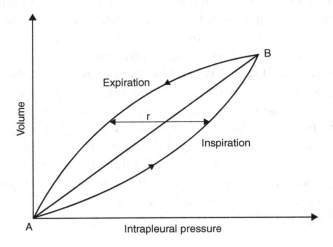

Fig. 44. Dynamic volume pressure loop obtained during breathing.

The Energy Needed to Breathe

To understand why the total work done during a respiratory cycle is given by the area enclosed by the dynamic volume pressure loop we will begin by thinking about the work done during inspiration:

$$E = \int Fdx$$

Now

$$F = \int PdA$$

so

$$E = \int \left(\int PdA \right) dx$$

and since

$$V = \int PdA$$

we have $E = \int PdV$, i.e. the area between the inspiratory volume-pressure curve and the y-axis in Fig. 45.

By the same argument, the elastic energy released during expiration is given by the area between the expiratory volume-pressure curve and the y-axis (area b). Much of the energy consumed during inspiration in doing elastic work is returned during expiration but clearly areas a and b are not equal: some energy must have been lost, i.e. net work must have been done. Most of the energy consumed as we breathe is used in doing work against the airway resistance (airway resistive work) and this accounts for most of area c. This is tantamount to saying that these loops also show a sort of *hysteresis* due to airways resistance. Incidentally, a little energy is also consumed doing work against frictional forces acting between tissues moving against each other (*tissue resistive work*): as mentioned earlier, this resistance is minimised by the presence of lubricating fluid in the intrapleural space.

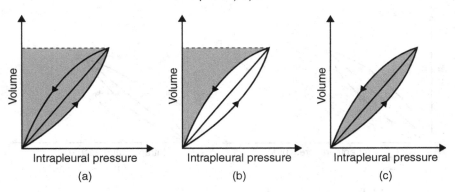

Fig. 45. Volume pressure loops.

Clinical Box 12: *Restrictive and Obstructive Lung Diseases*

In restrictive lung diseases such as *pulmonary fibrosis*, the average compliance of the lungs or chest wall during inspiration and expiration is decreased. This is illustrated in the volume-pressure loops shown in Fig. 46. For simplicity, these consider the lung tissue only and do not show hysteresis. Decreased average compliance is reflected in a decrease in the slope of the line AB. It follows that more work must then be done to cause air to flow into the lungs, either by achieving the same volume in each breath (by increasing the magnitude of the negative intrapleural pressure on inspiration and hence the area of the loop) or by increasing the number of breaths per unit time (by increasing the respiratory rate). Both these possibilities result in an increase in the work done per unit time (power output).

In obstructive lung diseases such as asthma, the resistance of the airways is increased, resulting in an increase in the horizontal width of the loop, *r*. In order to cause air to flow into the lungs the area of the loop must again be increased, representing an increase in the work done.

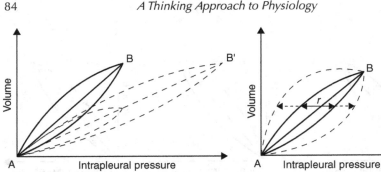

Fig. 46. Volume pressure loops in restrictive and obstructive lung diseases (dotted lines).

Clinical Box 12 where we consider the *restrictive* and *obstructive lung diseases* , clinical situations in which these factors change.

Ordinarily, the energy required for respiration represents something between 2% and 5% of the body's total energy consumption. Since at steady state our metabolism must be entirely aerobic this means that breathing accounts for 2% to 5% of O_2 consumption. In severe restrictive or obstructive lung disease, this may increase to 20% or more. Very sick patients may find themselves in an unwinnable positive-feedback cycle where O_2 is made available to the body at an insufficient rate, but increasing the depth or rate of breathing only serves to worsen the problem.

4.1.6 Elastic recoil and expiratory flow

A great deal of physiologically and clinically important information can be gathered by analysing relationships between air flow and lung volume. Such a curve plotted in a healthy patient is shown in Fig. 47. Flow reaches a peak soon after expiration begins and then just as quickly begins to decline. While the ascending portion of the curve is affected by effort, the descending portion of the curve is not. This reveals an interesting and clinically important property of the airways. It is the *longitudinal* pressure difference between the alveoli and the atmosphere that results in air flow out of the lungs during

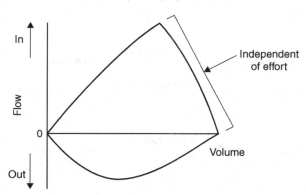

In

Flow

0

Out

Volume

Independent
of effort

Fig. 47. Flow volume loop.

expiration. However, it is the *transverse* pressure difference between the alveoli and the intrapleural space that results in the airways remaining open.

During forced expiration, the intrapleural pressure increases and becomes positive and therefore the intra-alveolar pressure also increases. An added contribution from the elastic recoil of the lung in fact elevates the intra-alveolar pressure above the intrapleural pressure: the transverse pressure difference between the intra-alveolar space and the intrapleural space is therefore positive. It is this positive pressure difference that keeps the airways open. As we move along the airways towards the mouth, the intra-airway pressure progressively falls because of airway resistance. At some point along the airways, intra-airway pressure will equal intrapleural pressure which is constant throughout the intrapleural space: this is known as the *equal-pressure point* (EPP). On quiet expiration, the EPP is located in the main bronchi which have the benefit of cartilaginous support to hold them open. However, on forced expiration, the EPP moves towards the lungs. If the EPP is reached in the airways distal to the terminal bronchioles, which lack support from cartilage, the airways will collapse and expiratory flow will be limited. Importantly, this means that it is the elastic recoil of the lung, determined by its compliance, that in fact determines maximum expiratory flow. Furthermore, expiratory flow cannot be increased by increasing effort. This is because increasing

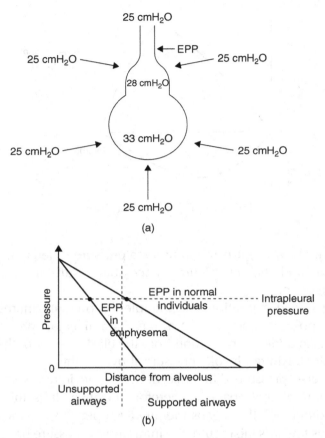

Fig. 48. Airway compression and the equal pressure point.

effort increases intrapleural pressure and an increase in intrapleural pressure will also proportionately increase the intra-alveolar pressure, thus having no effect on the EPP. This is illustrated in Fig. 48.

4.2   The Exchange of Gases Between the Lungs and the Blood

Having considered the process of moving O_2 and CO_2 between the atmosphere and the alveoli, we are now in a position to consider how these gases are exchanged between the alveoli and the blood. Naturally both these processes occur down an energy gradient. While

Clinical Box 13: *Chronic Obstructive Pulmonary Disease (COPD)*

The mechanical changes underlying COPD are somewhat more complex and interesting. COPD has two components: chronic bronchitis (clinically defined as a cough producing sputum for at least three months of the year for at least two consecutive years), which contributes an obstructive component, and emphysema (abnormal permanent dilatation of air airways distal to the terminal bronchioles). In emphysema, damage to the alveolar walls results in increased lung compliance. This is clearly not a problem during inspiration. However, it means that less elastic potential energy is then stored to be released on expiration. This leads to elastic recoil making a smaller-than-usual contribution to the intra-alveolar pressure. The transverse pressure gradient between the airways and the intrapleural spaces decreases, moving the EPP towards the lungs and causing the airways to collapse on expiration. This explains why patients with emphysema find some relief by breathing through pursed lips: breathing through a narrow tube increases airway resistance, decreasing flow and, as a consequence of Darcy's law, increasing the intra-airway pressure. This moves the EPP towards the mouth, hopefully to parts of the airways which are held open with cartilage and are therefore non-collapsible. The same reasoning explains why babies with IRDS grunt: without surfactant, alveolar surface tension is high, tending to make them collapse on expiration. Narrowing the vocal cords increases the resistance of the larynx and increases the intra-airway pressure. This moves the EPP upwards and helps to prevent the alveoli from collapsing.

a pressure gradient drives exchange between the atmosphere and the alveoli, a concentration gradient drives exchange between the alveoli and blood. Here, a concentration difference, rather than pressure difference, is key as gases are dissolved in solution as they enter the blood. However, pressures, or more specifically partial pressures, are still important as they in turn determine the concentration of a gas in solution. Here:

$$P_p = \text{fraction of gas mixture} \cdot \text{total pressure}$$

where P_p = partial pressure (Pa **or** mmHg). Hence

$$C = P_p \cdot S$$

where C = concentration ($mol \cdot m^{-3}$ **or** $mol \cdot L^{-1}$)
S = solubility in blood at body temperature ((m^3)$\cdot m^{-2} \cdot kg \cdot s^{-2}$
or $ml \cdot L^{-1} \cdot mmHg^{-1}$).

As with any flow, the *rate* at which this transfer occurs depends not only on this energy difference but also on the resistance to flow. Fick's law states that

$$\dot{j} = \frac{A}{d} \cdot D \cdot S \cdot \Delta P_p$$

where $\dot{j}$ = rate of diffusion ($m \cdot s^{-1}$)
A = area (m^2)
d = distance over which diffusion occurs (m)
D = constant of proportionality related to the diffusion co-efficient ($m \cdot (m^{-3}) \cdot s^3 \cdot kg^{-2}$ **or** $L \cdot (ml)^{-1} \cdot s^{-1}$)
ΔP_p = partial pressure difference (Pa **or** mmHg)

From this, we can see that the flow of gas between the lungs and the blood will depend on:

(1) the surface area of the alveoli over which exchange can occur (A);
(2) the combined distance across the alveolar membrane, the interstitial space and the capillary membrane (d);
(3) the diffusion coefficient for the gas (related to D);
(4) the solubility of each gas in blood (S);
(5) the difference in partial pressure between the alveoli and the blood for each gas (ΔP_p).

The first two factors apply equally to any gas while the final three are specific to a particular gas.

• *Surface area (A)*

The alveolar surface area is significantly larger than that of any other membrane in the body, averaging 75 m^2 (the area of a tennis court) in an adult man. Not all this area normally receives ventilation and perfusion, however. Increases in blood supply and ventilation as occur during exercise can therefore increase the area used and thus the rate of diffusion.

• *Diffusion distance (d)*

Despite the several layers separating alveolar gas and blood, diffusion distance is ordinarily small (around 0.3 μm). However, this may be increased if the alveolar membrane is thickened as in pulmonary fibrosis, or if fluid accumulates in the alveoli as in pulmonary oedema or pneumonia.

• *Diffusion coefficient (D)*

The diffusion coefficient for a gas is related to its molecular weight:

$$D \propto \frac{1}{\sqrt{MW}}$$

Since

$$\sqrt{\frac{MW_{CO_2}}{MW_{O_2}}} = \sqrt{\frac{32}{44}} = \sqrt{0.73} = 0.85$$

all other factors being equal, CO_2 should diffuse at 0.85 times the rate at which O_2 diffuses. However, this is massively offset by CO_2 being so much more soluble in water than O_2 as we will see below.

• *Solubility of each gas in blood (S)*

The major constituents of atmospheric gas are nitrogen, oxygen and carbon dioxide. Nitrogen constitutes 78% of atmospheric gas but we can largely ignore it here as it is not used by the body and therefore no concentration gradient exists for its exchange. This itself illustrates an important principle: it is only because the body consumes O_2 and produces CO_2 that concentration gradients exist for these gases. If it were possible to temporarily stop cells from respiring, O_2 and CO_2 exchange would cease.

Importantly, the solubilities of these gases in blood, or to a first approximation in water, differ significantly:

$$S_{O_2} = 0.03 \cdot ml \cdot L^{-1} \cdot mmHg^{-1}$$

$$S_{CO_2} = 0.69 \cdot ml \cdot L^{-1} \cdot mmHg^{-1}$$

Therefore, at the same partial pressure, the concentration of CO_2 will be 23 times that of O_2. All other factors being equal, CO_2 should diffuse 23 times faster than O_2. This is partially offset by differences in the diffusion coefficient, but 0.85×23 still leaves us with CO_2 diffusing 20 times faster than O_2. Under normal conditions this is not of great importance as there is sufficient time available for diffusion of both gases to occur. However, if diffusion is compromised for any other reason, perhaps because mucus and secretions increase the thickness of the diffusion barrier as in pneumonia, the transfer of O_2 is affected more rapidly and to a much greater degree than that of CO_2. This can result in a decrease in P_{aO_2} without a concomitant increase in P_{aCO_2}. Such *hypoxia* (defined as a P_{aO_2} of less than 60 mmHg) without *hypercapnia* (defined as a P_{aCO_2} of greater than 45 mmHg) is termed Type 1 respiratory failure. As a brief aside, hypoxia with hypercapnia is termed Type 2 respiratory failure. This usually results from mechanical disorders.

- *Partial pressure difference* (ΔP_p)

Since the solubility of a given gas in blood is effectively constant, it is changes in partial pressure that are key to determining changes in the concentration of gas dissolved in blood. Given that the surface area over which diffusion can occur is so large and that sufficient time is usually available for diffusion, it would not be unreasonable to expect that equilibrium should be reached and arterial partial pressures should be equal to alveolar partial pressure, i.e. $P_{aO_2} = P_{AO_2}$. In fact, this does not quite happen: even in healthy patients P_{aO_2} (circa 95 mmHg) is always less than P_{AO_2} (circa 100 mmHg). The reasons for this are worth some attention.

4.2.1 Ventilation and perfusion

There are two main reasons for this difference. The first is venous to arterial *shunting*. This describes the mixing of systemic venous blood that has not passed through the pulmonary circulation with systemic arterial blood. Such shunts may be either *anatomical* or *functional* (physiological).

In healthy subjects, there are two main causes of *anatomical* shunting: some coronary venous blood drains into the left ventricle via the Thebesian veins and some blood from the bronchial circulation drains into the pulmonary veins. Combined, these normally constitute less than 2% of the cardiac output. However, this can increase significantly when structural abnormalities of the heart cause blood to pass from its right side (deoxygenated) directly to its left side (oxygenated). This type of shunting can be identified clinically by the fact that breathing O_2 at an increased partial pressure does not improve P_{aO_2}.

Functional shunting also occurs in healthy patients. This is seen when alveoli are not ventilated but are still perfused, perhaps as a result of collapse or blockage with mucus. This leads us to the other cause of the difference between arterial and alveolar partial pressures. When standing upright the weight of the lungs tends to pull them towards to the chest wall to a greater degree at the base than at the apex. This results in the intrapleural pressure at the end of expiration being less negative at the base ($-2.5\,cmH_2O$) than at the apex ($-10\,cmH_2O$). The alveoli at the base of the lung are, therefore, not as expanded as those at the apex. Since lung compliance increases as volume decreases (see Fig. 44), the base has a higher compliance than the apex. The circa $3\,cmH_2O$ change in intrapleural pressure that occurs on inspiration will therefore lead to a greater change in volume at the base than at the apex, meaning that the base will be better ventilated. In fact when standing upright, the alveoli at the base of the lungs receive around 2.5 times as much ventilation per unit volume than alveoli at the apex. Notably the effect of such functional shunting on P_{aO_2} can be offset by breathing O_2 at an increased partial pressure.

As well as affecting alveolar ventilation, gravity also has marked effects on alveolar perfusion. While it has no influence on the extravascular (intra-alveolar) pressure it is important in determining the intravascular (hydrostatic) pressure. When standing upright, the base of the lung is around 10 cm below the tricuspid valve where blood leaves the heart. Assuming that the density of blood is the same as that of water, this means that the intravascular pressure will be increased by $10\,cmH_2O$ and this will in turn increase perfusion. By contrast, the apex is around $15\,cmH_2O$ above the tricuspid valve and so the intravascular pressure there will be decreased by $15\,cmH_2O$, decreasing perfusion. In fact, when standing upright alveoli at the base of the lungs receive around six times as much perfusion per unit volume as alveoli at the apex.

Despite both ventilation and perfusion being greater at the base than at the apex, the relative magnitudes of their values and the rates of decrease on moving up the lungs differ. This means that the ventilation to perfusion ratio ($V_A : Q_A$) varies throughout the lung (Fig. 49).

Let us imagine what would happen with extreme values of $V_A : Q_A$. If an alveolus is unventilated but well perfused, the partial pressures of gases in the blood leaving it will remain the same as those in the mixed-venous blood perfusing it. If on the other hand, an alveolus is well ventilated but completely unperfused, the partial pressures of gases in the alveolus will remain the same as in the inspired air as none of the oxygen will be able to enter the blood. Of course these extremes

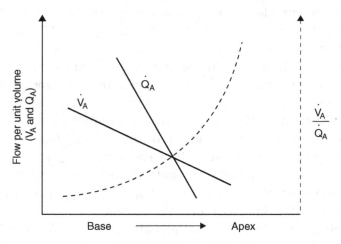

Fig. 49. Ventilation: perfusion relationships.

are not normally reached but they illustrate the principle at work. In a healthy lung, $V_A : Q_A$ is around 0.6 at the base while at the apex it is around 3. Interestingly *Mycobacteria*, which cause tuberculosis, grow best where P_{O_2} is high. This might explain why the apices are often affected first in pulmonary tuberculosis. For the lung as a whole, V_A is typically 5 L·min^{-1} while Q_A is typically 6 L·min^{-1}, giving an average $V_A : Q_A$ for the entire lung of 0.8. This results in an average P_{AO_2} of 100 mmHg and P_{ACO_2} of 40 mmHg. Since both V_A and Q_A are greater at the base than at the apex, the base makes a proportionally greater contribution to both the mix of expired gas and arterial partial pressures.

4.3 The Carriage of Gases in the Blood

Having considered how gases are transported into and out of the lungs and transferred to the blood, we now need to think about how they are carried in the blood. Given that 0.03 ml of O_2 dissolves in each litre of blood for each mmHg pressure, at a P_{AO_2} of 100 mmHg each litre of blood contains 3 ml of O_2. The body's resting O_2 consumption is around 300 ml·min^{-1} so if O_2 were exclusively carried dissolved in solution a cardiac output of 100 L·min^{-1} would be needed. Clearly, this is wildly unrealistic so some other means of transporting O_2 is absolutely essential. Up to 98.5% of O_2 in blood is carried bound to haemoglobin (Hb) molecules inside red blood cells. Binding to this protein increases the total O_2 content of each litre of blood to around 200 ml, compatible with a far more reasonable cardiac output. Happily, while Hb dramatically increases the O_2-carrying capacity of the blood, the binding of O_2 to Hb has no effect on either the concentration of O_2 dissolved or on the P_{aO_2} so everything said above still holds true.

We need not concern ourselves here with the biochemistry of the Hb molecule but we should think a little about its properties.

The total amount of O_2 bound to Hb per unit volume of blood must depend on the number of Hb molecules per unit volume of blood, the ability of these molecules to bind O_2 (their *affinity*) and the proportion of Hb molecules that have bound O_2 (are saturated with O_2).

It is intuitive that the more Hb molecules per unit volume of blood, the greater the O_2 content will be. Conversely, if the number of Hb molecules per unit volume of blood decreases, the capacity for O_2 carriage will decrease. Decreases in the Hb content of the blood or in the ability of Hb to bind O_2 occur in *anaemia*. Since in either case, there is in effect less useful Hb available, the total O_2 content of blood at any given P_{aO_2} will be decreased. If the circulatory system is to continue to deliver the same quantity of O_2 per unit time to the tissues, then the cardiac output must increase. This explains why anaemia is particularly associated with an increase in heart rate.

The proportion of these Hb molecules saturated with O_2 depends on the P_{aO_2}. This relationship is described by the *O_2-Hb dissociation curve* (which could just as well be called an *association* curve) (Fig. 50). The characteristic sigmoidal shape of this curve results from each Hb molecule being able to bind four O_2 molecules in a *cooperative* manner so that the affinity of Hb for O_2 increases as each of four O_2 molecule binds. The curve is flat over the normal working range of P_{AO_2} values, meaning that blood still becomes almost fully saturated

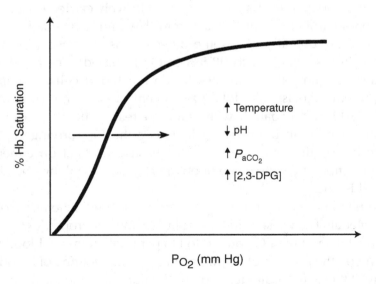

Fig. 50. O_2-haemoglobin dissociation curve.

with O_2 as it passes through the lungs. Notably, P_{aO_2} values as low as 50 mmHg have been reported in healthy subjects during a long haul flight. Nevertheless, this is usually still on the flat part of the curve and therefore Hb saturation may still be greater than 90%. As helpful as this seems, it also means that manoeuvres such as hyperventilation or breathing O_2 at a high partial pressure do not significantly increase the O_2 content of the blood in a healthy person. This also necessitates caution when interpreting *pulse oximeter* readings. These give a simple clinical indication of a patient's Hb saturation. While such readings are useful in identifying when P_{aO_2} is very low, they do not reflect mild or moderate decreases in P_{aO_2} as large decreases are needed before Hb saturation starts to fall.

The shallow slope of the O_2-Hb dissociation curve over the values of P_{O_2} encountered in the lungs limits the effects of small changes in P_{aO_2} on the O_2 content of the blood. In contrast, its steep slope over the values of P_{aO_2} normally seen at the tissues allows Hb to give up its bound O_2. What is more, a small decrease in tissue P_{aO_2} results in a significant decrease in Hb saturation, i.e. increases unloading. More active tissues with lower P_{aO_2} values will therefore receive more O_2 than resting tissues with higher P_{aO_2}s. Active tissues also produce more CO_2 and H^+. Both of these decrease the affinity of the Hb for O_2, shifting the dissociation curve to the right (see Fig. 50) and further facilitate O_2 unloading. This is referred to as the *Bohr effect*. In addition, the rate of production of 2,3-diphosphoglycerate (2,3-DPG, a sugar) increases as the rate of glycolysis increases. This acts directly on the Hb molecule to decrease O_2 affinity and further shift the curve to the right. Together, these effects mean that the amount of O_2 unloaded in the tissues is primarily dependent on the activity, and therefore demand, of the tissues.

In contrast to O_2, CO_2 is readily soluble in water. While less than 1.5% of O_2 carried in blood is dissolved in solution, up to 6% of the total CO_2 content of blood is represented by dissolved gas. Of the remaining 94%, around 5% is carried as carbamino compounds, formed when CO_2 binds to amine groups on plasma proteins and Hb. The most important form in which CO_2 is carried, however, is HCO_3^-. When dissolved in plasma CO_2 reacts with water to form carbonic

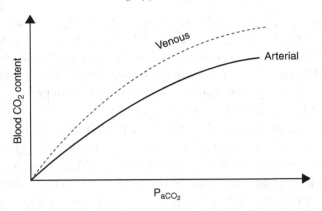

Fig. 51. Relationship between P_{aCO_2} and blood CO_2 content.

acid:

$$CO_2 + H_2O \rightleftharpoons H_2CO_3$$

As H_2CO_3 is a weak acid, it does not significantly dissociate. However, it is able to diffuse into red blood cells and these contain a vital enzyme, *carbonic anhydrase* (CA) which dramatically increases the rate at which carbonic acid dissociates:

$$H_2CO_3 \underset{}{\overset{CA}{\rightleftharpoons}} H^+ + HCO_3^-$$

H^+ can then be buffered inside the red blood cell. As explained above, as Hb binds H^+, its affinity for O_2 falls (the *Bohr effect*), encouraging Hb to give up O_2 at the tissues where the P_{CO_2} is high. The counterpoint is also true: the affinity of Hb for H^+ is increased as O_2 unloads (*the Haldane effect*). The relationship between blood CO_2 content and P_{aCO_2} is shown in Fig. 51.

HCO$_3^-$ is exchanged across the cell membrane for Cl$^-$ (the chloride shift or *Hamburger phenomenon*). This maintains the intracellular charge, and hence the potential difference across the cell membrane, while allowing HCO$_3^-$ to enter the plasma and perform its vital function as a buffer (see later). This in turn results in net accumulation of Cl$^-$ which this leads to an osmotic gradient for water entry into the red blood cell. Increased red blood cell volume as a result of CO_2-loading explains why the proportion of the blood volume accounted for by

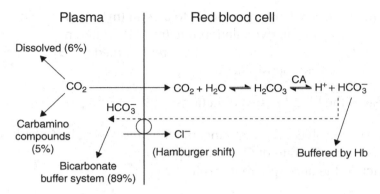

Fig. 52. Summary of CO_2 handling.

red blood cells (the packed red cell volume) is some 3% larger in the systemic veins as compared to the arteries! These processes are summarised in Fig. 52.

FURTHER THOUGHTS

4.4 The Alveolar Gas Equation

It is clear that the delivery of O_2 to the tissues is ultimately dependent on maintaining a sufficient partial pressure of O_2 in the alveoli. With a few assumptions, we can derive a simple equation describing the key factors determining the partial pressure exerted by O_2 in the alveoli. We will go on to make the assumption throughout that the partial pressures of gases in the alveoli are approximately equal to those in the arterial blood (P_{aO_2}). The same applies for CO_2. We will begin by considering the inputs and outputs of gases to and from the alveoli.

O_2 enters the alveoli from the atmosphere at a rate which depends on the volume of gas entering the alveoli per unit time, the alveolar ventilation rate.

$$\text{input} = \dot{V}_A \cdot F_{IO_2}$$

where input = rate of input of O_2 to alveoli ($m^3 \cdot s^{-1}$ **or** $L \cdot min^{-1}$)

V_A = alveolar ventilation rate ($m^3 \cdot s^{-1}$ **or** $L \cdot min^{-1}$)

F_{IO_2} = fraction of inspired gas represented by O_2
(dimensionless)

O_2 leaves the blood in two directions:

- across the alveolar membrane into the blood (which at steady state represents the rate of O_2 consumption).
- back to the atmosphere in expired gas.

The first of these is simply given by the rate of O_2 consumption. The second is given by $V_A \cdot F_{AO_2}$:

$$\text{output} = \dot{V}_{O_2} + \dot{V}_A \cdot F_{AO_2}$$

where output = rate of output of O_2 from alveoli ($m^3 \cdot s^{-1}$ **or** $L \cdot min^{-1}$)

F_{AO_2} = fraction of alveolar gas represented by O_2
(dimensionless)

If the P_{AO_2} is constant, then the rate at which O_2 is being added to the alveoli (input) must be equal to the rate at which it is removed (output). Therefore:

$$\dot{V}_A \cdot F_{IO_2} = \dot{V}_{O_2} + \dot{V}_A \cdot F_{AO_2}$$

rearranging

$$\dot{V}_{O_2} = \dot{V}_A (F_{IO_2} - F_{AO_2})$$

4.4.1 Factors determining P_{AO_2}

If we assume that the alveolar pressure is equal to the atmospheric pressure and that the air is dry, then by Dalton's law:

$$F_{O_2} = \frac{P_{O_2}}{P_{atmos}}$$

where F_{O_2} = fraction of gas represented by O_2 (dimensionless)
P_{O_2} = partial pressure of O_2 (Pa **or** mmHg)
P_{atmos} = atmospheric pressure (Pa **or** mmHg)

so

$$\dot{V}_{O_2} = \dot{V}_A \left(\frac{P_{IO_2} - P_{AO_2}}{P_{atmos}} \right),$$

where P_{IO_2} = partial pressure of O_2 in inspired gas (Pa **or** mmHg)
P_{AO_2} = partial pressure of O_2 in alveolar gas (Pa **or** mmHg)

which rearranges to:

$$P_{AO_2} = P_{IO_2} - \frac{\dot{V}_{O_2} \cdot P_{atmos}}{\dot{V}_A}.$$

This allows us to examine graphically the effect of each parameter on P_{AO_2}. To keep things simple, we plot P_{AO_2} against $\dot{V}_A$, changing each of the other parameters one at a time (Figs. 53 and 54).

For each of the family of hyperbolae, P_{AO_2} can be increased by increasing $\dot{V}_A$, i.e. by hyperventilating, and decreased by decreasing $\dot{V}_A$, i.e. by hypoventilating. From the equation, increasing P_{IO_2} shifts the curve upwards (Fig. 53, arrow), while decreasing P_{IO_2} shifts the curve downwards. In a mechanically ventilated patient, it would be

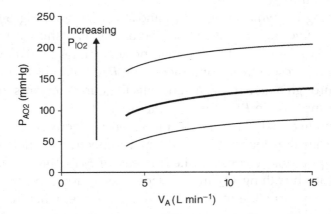

Fig. 53. The effect of increasing P_{IO_2} on the relationship between P_{AO_2} and $\dot{V}_A$.

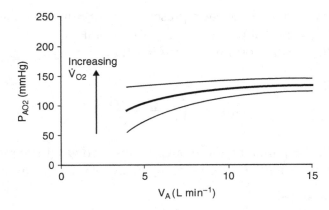

Fig. 54. The effect of decreasing V_{O_2} on the relationship between P_{AO_2} and V_A.

possible to alter P_{atmos} while keeping P_{IO_2} and V_A constant, but in reality, this situation never arises. As can be seen from the equation, increasing P_{atmos} would be expected to shift the curve downwards, while decreasing P_{atmos} would be expected to shift it upwards. Importantly, a high altitudes both P_{IO_2} and P_{atmos} are decreased (F_{IO_2} is constant). The combined effect of these changes is to shift the curve relating P_{AO_2} to V_A downwards, necessitating an increased V_A to maintain the P_{AO_2}.

Increasing V_{O_2} while holding P_{IO_2} and P_{atmos} constant (Fig. 54, arrow) shifts the curve downwards while decreasing V_{O_2} shifts it upwards. Keeping P_{atmos} and V_{O_2} fixed at their normal values (760 mmHg and 300 ml min^{-1} respectively) and increasing P_{IO_2}, perhaps by breathing oxygen through a face mask, again shifts the curve upwards, resulting in a large increase in P_{AO_2} for all values of V_A.

It is intuitive that P_{AO_2} can be increased by increasing V_A but as we see from the equation, it is far more effective to increase P_{IO_2} while keeping P_{atmos} constant, i.e. increasing F_{IO_2}. This is effectively achieved by breathing O_2 from a face mask. However, this option is not usually available in everyday life. Therefore, when faced with an increased O_2 demand, increasing respiratory rate and volume is the most effective option available. P_{AO_2} can also be increased by decreasing V_{O_2}. Of course P_{AO_2} could also be increased by decreasing P_{atmos} while keeping P_{IO_2} constant, again increasing F_{IO_2}, but this situation seldom exists.

4.4.2 Factors determining P_{ACO_2}

Now let us consider CO_2. Assuming that the tissues do not accumulate CO_2, it must enter the circulation at the rate at which it is produced by the tissues (around $200\,ml \cdot min^{-1}$). Let us assume that all CO_2 produced is immediately delivered to the alveoli, i.e. diffusion across the alveolar capillary membranes occurs completely and instantaneously.

$$\text{input} = \dot{V}_{CO_2}$$

where input = rate at input of CO_2 to alveoli ($m^3 \cdot s^{-1}$ **or** $L \cdot min^{-1}$)
$\dot{V}_{CO_2}$ = rate of CO_2 production ($m^3 \cdot s^{-1}$ **or** $L \cdot min^{-1}$)

CO_2 leaves the alveoli for the atmosphere at a rate dependent on alveolar ventilation:

$$\text{output} = \dot{V}_A \cdot F_{ACO_2}$$

where output = rate of output of O_2 from alveoli ($m^3 \cdot s^{-1}$ **or** $L \cdot min^{-1}$)
F_{ACO_2} = fraction of alveolar gas represented by CO_2
(dimensionless)

If P_{ACO_2} is constant then V_{CO_2} (input) must be equal to the rate at which it is being removed (output). Therefore:

$$V_{CO_2} = \dot{V}_A \cdot F_{ACO_2}$$

By Dalton's law:

$$F_{ACO_2} = \frac{P_{ACO_2}}{P_{atmos}}$$

where P_{ACO_2} = partial pressure of CO_2 in alveolar gas (Pa **or** mmHg).

So:

$$\dot{V}_{CO_2} = \frac{P_{ACO_2}}{P_{atmos}} \cdot \dot{V}_A$$

and this rearranges to

$$P_{ACO_2} = \frac{\dot{V}_{CO_2} \cdot P_{atmos}}{\dot{V}_A}$$

and allows us to plot a family of curves this time relating P_{ACO_2} to V_A for different values of V_{CO_2} (Fig. 55). By inspection, P_{ACO_2} is inversely proportional to V_A and so have formally shown that the more we breathe, the more CO_2 we expire and the lower the P_{ACO_2}. Since P_{ACO_2} is the key factor determining V_A (providing our drive to breathe), this constitutes a hugely important example of negative feedback.

Of course, this assumes that the inspired air contains no CO_2. This is usually a fair assumption: in atmospheric gas P_{CO_2} is 0.21 mmHg, i.e. negligible. However, if a subject breathes a gas mixture containing a significant P_{CO_2} then the expression becomes:

$$P_{ACO_2} = \frac{V_{CO_2} \cdot P_{atmos}}{\dot{V}_A} + P_{ICO_2}$$

where P_{ICO_2} = partial pressure of CO_2 in inspired gas (Pa **or** mmHg) and this shifts the curve upwards along the y-axis (Fig. 56).

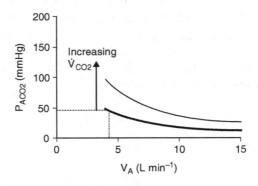

Fig. 55. The effect of increasing V_{CO_2} on the relationship between P_{ACO_2} and V_A (dotted lines indicate normal values).

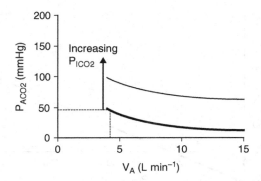

Fig. 56. The effect of increasing P_{ICO_2} on the relationship between P_{ACO_2} and V_A (dotted lines indicate normal values).

4.4.3 The alveolar gas equation

We can now think about combining these two equations. Taking a step aside, the respiratory quotient is defined as:

$$R = \frac{\dot{V}_{CO_2}}{\dot{V}_{O_2}}$$

where R = respiratory quotient (dimensionless).

So

$$R = \frac{\dot{V}_A \cdot F_{ACO_2}}{\dot{V}_A(F_{IO_2} - F_{AO_2})}$$

which simplifies to

$$R = \frac{F_{ACO_2}}{F_{IO_2} - F_{AO_2}}$$

and rearranges to

$$F_{AO_2} = F_{IO_2} - \frac{F_{ACO_2}}{R}$$

or incorporating Dalton's law

$$P_{AO_2} = P_{IO_2} - \frac{P_{ACO_2}}{R}$$

both of which are forms of the *alveolar gas equation*.

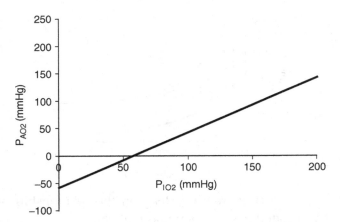

Fig. 57. The relationship between P_{AO_2} and P_{IO_2} as described by the alveolar gas equation.

It must be noted that this equation does not make allowance for the contribution of water vapour to the alveolar gas mixture. At body temperature, water vapour contributes a partial pressure of 47 mmHg, so if P_{atmos} is 760 mmHg the true pressure being contributed by the gases is closer to 713 mmHg.

The equation we have derived also makes at least one nonsensical prediction. Consider the relationship between P_{AO_2} and P_{IO_2} (Fig. 57). It does not seem likely that P_{AO_2} will ever become negative! If the input of O_2 to the circulation is not sufficient to match the output from the circulation, then clearly, the equation becomes invalid and no longer represents the physiology. The P_{IO_2} at the top of Everest is around 52 mmHg and by this point this simple form of the alveolar gas no longer holds true.

One further point is worth noting. Throughout the derivation of the alveolar gas equation, we implicitly assume that we are working at a fixed lung volume, say the volume at the end of expiration. In reality, R is always less than 1, i.e. V_{CO_2} is less than V_{O_2}. Therefore the amount of matter in the lungs must be constantly decreasing (less CO_2 is being added than O_2 is being removed). Since pressure is directly proportional to the amount of matter present (by the ideal gas law), this must mean that intra-alevolar pressure is decreasing. This generates an additional pressure gradient for air to enter the lungs. This gradient is present at all times, even at the end of expiration, and serves to stabilise lung volume. This also has the effect of boosting the F_{AO_2}, and therefore the P_{AO_2}. It is possible to derive a somewhat more complex form of the alveolar gas equation which takes account of this.

This passive inflow of gas is not just a physiological curiosity but can be of clinical importance. When O_2 consumption is not too high, it is possible to maintain P_{aO_2} at an adequate level by administering pure O_2 via a face mask, even if the patient is not breathing! In fact, if gas exchange is normal and the patient is fully-oxygenated at the outset, it is the accumulation of CO_2, rather than a lack of O_2, which limits this type of ventilation. This in itself can be rather useful in patients recovering from anaesthesia: the increasing P_{aCO_2} stimulates breathing.

CHAPTER 5

The Circulatory System

All living organisms need to be able to transport nutrients and waste products to and from their cells. Small multicellular organisms are able to rely on simple diffusion for this purpose, while larger organisms have developed circulatory systems to overcome the increased distances between cells. Central to such circulatory systems is *blood*, a connective tissue of liquid and cells that acts as the medium for transport. This is then transported around the body in a closed system. In brief, blood travels away from the heart in *arteries*. These divide into *arterioles* and then *capillaries*, which have a large surface area to exchange substances with cells. Blood then enters *venules* before passing into *veins* and returning to the heart.

In humans, there are two functionally distinct circulatory systems (*circulations*). The *systemic circulation* delivers oxygenated blood from the left ventricle to the tissues and returns deoxygenated blood from the tissues to the right atrium. The *pulmonary circulation* delivers deoxygenated blood from the right ventricle to the lungs and returns oxygenated blood back to the left atrium (see Fig. 58). Although there are some important differences, the same principles apply to both circulations. We will first consider the systemic circulation then return to the pulmonary circulation.

5.1 Flow in the Circulatory System

Just as Ohm's law defines the flow of charge along a wire, Darcy's law defines the flow of volume along a pipe. As with most laws

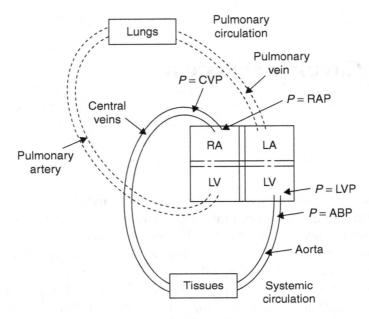

Fig. 58. Organisation of the circulatory system.

in physics, there are several ideal conditions which must be met (assumptions that must apply) in order for Darcy's law to be valid: these are dealt with in Further Thoughts. Assuming these assumptions to be true, then:

$$\Delta P = \dot{V} \cdot R_{\text{fluid}}$$

where ΔP = pressure difference driving flow (Pa **or** mmHg)
$\dot{V}$ = flow along the tube (m$^3 \cdot$s^{-1} **or** L$\cdot$min^{-1})
R_{fluid} = resistance to fluid flow (kg$\cdot$m$^{-2} \cdot$s^{-2} **or** mmHg$\cdot$L$^{-1} \cdot$min).

Blood will move around the circulatory system whenever there is a pressure difference to drive flow. In the arteries, the heart generates this pressure difference. It should be noted from the outset that the heart is *not* responsible for generating the pressure difference needed to drive flow in the veins: the heart pushes but does not suck. We will return to this issue later.

We can now begin our discussion by considering pressures at three points in the system: the pressure at which blood leaves the left ventricle, the pressure at the tissues and the pressure at which blood returns to the right atrium.

5.1.1 The pressure as blood leaves the left ventricle

Blood leaves the left ventricle to enter the aorta, which then branches to supply the peripheral arteries. The arteries offer very little resistance to flow. It thus follows from Darcy's law that there is only a very small pressure drop between the aorta and the peripheral arteries. We can therefore approximate the *left ventricular pressure* (LVP) as the *arterial blood pressure* (ABP).

5.1.2 The pressure at the tissues

The mean pressure at the venous side of tissues is referred to as the *mean systemic filling pressure* (MSFP). MSFP is a rather odd concept and can be thought of in a number of ways. It is defined as the average pressure that would exist in the systemic circulation were the heart to be stopped and the blood distributed among all vessels in proportion to the relative capacities of vessels. It therefore critically depends on the circulating blood volume. Although this is a rather abstract concept, MSFP is nonetheless a useful idea as we will see later. MSFP is around 7 mmHg in a healthy subject at rest.

5.1.3 The pressure at which blood returns to the right atrium

Blood returns to the heart at the right atrium and the pressure at the point of arrival is the *right atrial pressure* (RAP). Just as the LVP is reflected by the ABP, the RAP is reflected by the *central venous pressure* (CVP). It is conventional to discuss ABP and RAP and we will not depart from this here.

Pulsatile Flow in the Arteries

The ventricles sequentially contract (*systole*) and relax (*diastole*). The ventricular pressure must therefore also alternate from a maximum value during systole to a minimum during diastole. They can thus be thought of as behaving like the alternating potential difference producing an alternating current (a.c., Fig. 59(a)).

For blood to flow from the left ventricle into the aorta, the left ventricular pressure must exceed aortic pressure. This is the case during systole. However, during diastole aortic pressure exceeds left ventricular pressure. Were it not for the aortic valve, which allows flow only in the forward direction, this would result in back-flow from the aorta into the left ventricle. This situation may arise when the aortic valve is diseased, resulting in the regurgitant back-flow of blood into the left ventricle during diastole (aortic *regurgitation*). While from Darcy's law, we would expect flow to instantaneously match the pressure gradient, the aortic valve ensures that flow in the aorta occurs only in the forward direction: continuing an electrical analogy, the aortic valve behaves as *diode*.

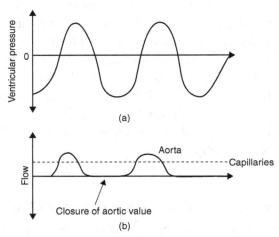

Fig. 59.   Flow in the aorta as compared to flow in the capillaries.

(Continued)

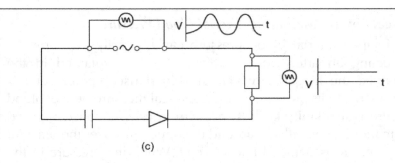

(c)

Fig. 59.   Pulsatile and non-pulsatile flow.

When applying Darcy's law to the arterial system, we usually get around the problem of aortic pressure changing with time by calculating an average value. At rest, the heart spends around two thirds the time in diastole and one third in systole. Thus:

$$\overline{ABP} = \frac{2}{3} \text{ diastolic pressure} + \frac{1}{3} \text{ systolic pressure}$$

where $\overline{ABP}$ = mean arterial blood pressure (Pa **or** mmHg).

Notably, by the time blood reaches the capillaries and venules, flow really is constant, rather like direct current (d.c.), giving a horizontal line on our plot. Returning to the electrical analogy, to convert (*rectify*) a.c. into d.c. in the simplest possible way two components are needed, a diode to allow flow in only one direction and a capacitor to smooth peaks and troughs. By analogy, Fig. 59(c) illustrates the simplest rectifying electrical circuit, termed a *half-wave rectifier*. It shows an alternating potential difference source viewed on an oscilloscope. The resulting alternating current is converted into a direct current by a diode and capacitor arranged in series. In the circulatory system, the aortic valve takes the role of the diode. Furthermore, the stretchy (compliant) walls of the elastic arteries such as the aorta take the role of the capacitor, stretching and storing blood during systole then releasing it during diastole.

Clinical Box 14: *Measuring Arterial Blood Pressure*

In principle, arterial blood pressure can be easily measured by connecting an artery to a vertical tube. As explained in the Appendix, the height to which the blood rises up the tube is proportional to its pressure. Routine clinical measurement of blood pressure is thankfully far less invasive. In brief, a cuff connected to a manometer is inflated around the upper arm over the brachial artery and air is pumped into the cuff. When the pressure in the cuff exceeds systolic ABP, flow ceases and the radial pulse can no longer be felt. The pressure in the cuff is then slowly lowered while listening over the brachial artery with a stethoscope. When the pressure in the cuff falls just below systolic ABP blood begins to flow and a dull tapping sound (*Korotkoff* sound) is heard. These sounds become louder and more murmuring as the cuff pressure if lowered further. As the cuff pressure is lowered further still these sounds abruptly disappear and this is taken as the diastolic ABP, though in fact, this method overestimates diastolic ABP by around 10 mmHg. The origin of Korotkoff sounds has long been a subject of debate. They have variously been attributed to the arterial wall suddenly stretching as the vessel refills with blood, to high flow velocities producing turbulence and to gas bubbles coming out of solution.

We are now almost ready to write simple equations for flow in the arterial and venous systems. First, however, we must discuss the final parameter in Darcy's law: resistance. The arterial circulation offers a high resistance to flow. This is referred to as the *arterial resistance* (ArtRes) and is anatomically located predominantly in the arterioles, narrow vessels arranged in series (see Appendix). In contrast, the venous circulation offers a negligible resistance to flow and it is not in fact possible to define the anatomical location of this resistance. *Resistance to venous return* (RVR) is therefore a theoretical

Clinical Box 15: *Measuring Central Venous Pressure*

Just as the direct measurement of ABP requires that a manometer be connected to an artery, the direct measurement of CVP requires that a monometer be connected to a large vein (usually the *jugular vein* in the neck). Such accurate measurements of CVP are sometimes necessary. However, a good enough impression of the CVP can usually be obtained simply by determining the height to the highest visible pulsation in the *internal jugular vein* above the right atrium (the *jugular venous pressure*, JVP). A convenient anatomical landmark can be used to locate the right atrium: in a subject lying down with back inclined at 45° to the horizontal, it is around 5 cm vertically below the angle formed by the junction of the manubrium and the sternum (*angle of Louis*). The JVP is therefore the vertical distance between a horizontal line running from 5 cm below the angle of Louis and a second horizontal line running from the highest point in the internal jugular vein at which pulsations can be see. The JVP is often less than 5 cm H_2O in healthy subjects and it may not, therefore, be possible to see pulsations.

conceptualisation of the resistance offered by the venous circulation. It is a rather strange concept: in fact, it is predominately not really a resistance at all, as we will see later. RVR is only really a useful concept in relation to Darcy's law. The total resistance of the systemic circulation (*total peripheral resistance*, TPR), encompasses the venous as well as the arterial circulations. However, since RVR is so small in comparison to ArtRes, TPR can usefully be thought of as the resistance offered by the arterial circulation.

Applying Darcy's law to the circuit shown in Fig. 58, we can write equations for the arterial and venous sides of the systemic circulation:

$$\overline{ABP} - MSFP = CO \cdot TPR \quad \text{and} \quad MSFP - RAP = VR \cdot RVR$$

where MSFP = mean systemic filling pressure (Pa **or** mmHg)
   CO = cardiac output ($m^3 \cdot s^{-1}$ **or** $L \cdot min^{-1}$)
   TPR = total peripheral resistance ($kg \cdot m^{-2} \cdot s^{-2}$ **or** $mmHg \cdot L^{-1} \cdot$
      min)
   RAP = right atrial pressure (Pa **or** mmHg)
   VR = venous return ($m^3 \cdot s^{-1}$ **or** $L \cdot min^{-1}$)
   RVR = resistance to venous return ($kg \cdot m^{-2} \cdot s^{-2}$ **or** $mmHg \cdot L^{-1} \cdot$
      min)

CO is defined as the volume of blood ejected from the left ventricle and VR as the volume of blood returning to the right atrium, per unit time (both around 5 $L \cdot min^{-1}$ at rest). This is often simplified to give an equation encompassing the entire system:

$$\overline{ABP} - RAP = CO \cdot TPR$$

However, it must be remembered that the heart is only responsible for generating flow around the arterial component. Here, CO really represents the average flow around the entire system rather than the volume of blood ejected from the left ventricle. Since RAP is usually close to zero we can approximate this equation by writing:

$$\overline{ABP} = CO \cdot TPR$$

While this chapter focuses on flow around the circulatory system as a whole, it is important to remember that these principles apply just as well to circulations in individual tissues such as skeletal muscle, brain or skin. Since in general the blood supply to tissues is arranged in parallel (see Appendix), each tissue is perfused with blood at a similar pressure, i.e. ABP is relatively constant. The body therefore has the capability to control blood flow through individual tissues by varying the resistances of their supplying vessels (see Chapter 7).

5.2   The Cardiac Cycle

Having established that it is the role of the heart to generate a pressure gradient sufficient to cause blood to flow along the arterial circulation, we can now begin to discuss the underlying sequence of events. It is most straightforward to describe the cardiac cycle as a pressure volume loop, just as we described the respiratory cycle as a volume pressure loop. Incidentally, which quantity goes on the x- and which on the y-axis does not matter, but conventionally volume-pressure loops are plotted for the respiratory cycle (see Chapter 4) and pressure-volume loops for the cardiac cycle. Here we plot the pressure in the left ventricle on the y-axis against the volume of blood in the left ventricle on the x-axis (Fig. 60). We could begin describing the loop at any point, but we will begin at **A**.

At **A**, the pressure in the left ventricle is less than the pressure in the left atrium. Consequently the mitral valve, which sits between the left atrium and left ventricle, is sucked open causing blood to flow from the left atrium into the left ventricle. During the initial phase of filling, ventricular pressure continues to fall as the muscle rapidly relaxes. Later in the filling phase, the curve plateaus as the ventricular muscle approaches the limit of the extent to which it can relax: ventricular pressure may even begin to increase near the end of the filling phase. Atrial contraction occurs near the end of the filling phase. It is important to note that this makes a maximum contribution of just 15% of ventricular filling, a point we will return to later.

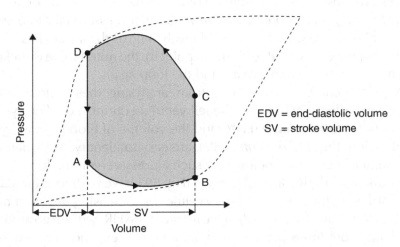

Fig. 60.   Pressure volume loop representing the cardiac cycle.

At point **B**, the pressure in the ventricle comes to exceed the pressure in the atrium (which by this point has fallen as the atrium empties and the ventricle fills), causing the mitral valve to close. The ventricle then begins to actively contract. However, both the inflow (mitral) and outflow (aortic) valves are closed, making the ventricle a sealed chamber from which blood cannot escape. Ventricular pressure therefore increases but volume does not change: this is referred to as the phase of *isovolumetric contraction*. When the ventricular pressure exceeds the aortic pressure (which at this point is low), a pressure gradient for flow exists and pushes the aortic valve open at point **C**. Blood flows down its pressure gradient and ventricular ejection begins. Despite the ensuing decrease in ventricular volume, ventricular pressure continues to increase as the ventricle still actively contracts and produces force. Later in this phase, the ventricle ceases to actively produce force and while blood continues to leave the ventricle down its pressure gradient, emptying decreases passive stretch and pressure begins to fall.

At point **D** ventricular pressure falls below aortic pressure, which is now higher since the aorta is filled with blood. This causes the aortic valve to close, preventing the back-flow of blood. The ventricle is again a sealed chamber and then begins to relax. Since both valves are closed, the ventricular volume is fixed: this is the phase of *isovolumetric relaxation*. When the ventricular pressure falls to below the atrial pressure (which is now relatively high as the atrium has been filling since the mitral valve closed at point B), the mitral valve is sucked open and we begin to move around the loop again.

A great deal can be learned from analysing these curves. The minimum volume of blood in the left ventricle during a cardiac cycle, the *end-diastolic volume* (EDV) and the volume of blood ejected per contraction, the *stroke volume* (SV), are easy to identify. As explained in Chapter 4, the area enclosed by such a pressure-volume loop gives the work done during a cardiac cycle (Fig. 60). If we wished to increase the cardiac output, we could do so either by increasing the heart rate (HR) or the SV (CO = HR · SV). If we increase the HR, we will go around the loop more times per unit time and thereby expend more energy per unit time, i.e. develop a greater power. To increase the stroke volume, we must shift point **B** to the right. This will have the effect of increasing the area enclosed by a single loop, increasing the work done per cardiac cycle and thereby the power developed. It is usually more efficient to increase cardiac output by increasing stroke volume than by increase heart rate, but in reality we usually do both.

Interestingly, increasing SV not only shifts point B to the right along the x-axis but also upwards along the y-axis. This can be explained by the passive properties of ventricular muscle: increasing ventricular volume increases the stretching force experienced by muscle fibres, i.e. it decreases the change in volume produced by a unit change in pressure (*compliance*, see Chapter 4). This in turn increases ventricular pressure. The gradient of the dashed line in Fig. 60 represents the reciprocal of compliance. This gradient increases, representing a decrease in compliance, as stretch increases. This makes sense, after all the more a rubber band is stretched the less stretchy it becomes.

5.3   Pre-Load and After-Load

It is now time to consider the mechanical factors which affect how hard the heart must work to pump blood. If we consider an isolated strip of cardiac muscle *in vitro*, we can think of two modes of loading. The muscle can be pre-stretched by a weight before it begins to contract (a *pre-load*) or made to stretch by an additional weight as it begins to contract and shorten (an *after-load*). These are illustrated in Fig. 61.

There is great debate about the most useful and meaningful ways to define these terms for the heart *in vivo*. Fundamentally, pre-load is the stretch on the ventricular muscle fibres before systolic contraction begins. A fair index of this is the volume of blood in the ventricle at this time, the EDV. Ultimately, this is determined by the volume of blood entering the ventricle during the filling phase, which is in turn determined by the atrial filling pressure. In contrast, after-load is the stretch on the ventricular muscle fibres during systole and is determined by all factors which oppose the movement of blood from the ventricle to the artery. A reasonable index of this is the systolic ABP, though other indices such as ArtRes are sometimes used: the former definition is most useful for our purposes.

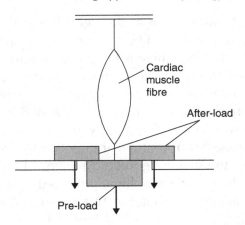

Cardiac
muscle
fibre

After-load

Pre-load

Fig. 61. Laboratory demonstration of preload and afterload.

If pre-load is increased, more blood enters the ventricles during each filling cycle (Fig. 62). This increases SV and shifts **B** to the right along the x-axis and up the y-axis a little. Isovolumetric contraction therefore begins at a higher volume and a slightly higher pressure. The ventricle must work harder to eject this larger volume of blood. The loop becomes wider horizontally but cannot become shorter vertically if all blood is still to be ejected. This means that the loop is stretched horizontally, increasing its area, reflecting an increase in the work done per cardiac cycle.

If after-load (ABP) is suddenly increased, a greater left ventricular pressure (LVP) is needed to cause the aortic valve to remain open and blood to flow into the aorta (Fig. 63). Moving around the loop, the aortic valve will similarly close at a higher pressure as the relative pressure difference resulting in closure will be the same. This will shift **D** upwards. The result is that isovolumetic relaxation begins at a larger volume and blood backs-up in the ventricle. This increases the stretching forces experienced by the muscle cells (effectively increasing pre-load) and, by Starling's law (see below), results in an increase in energy of contraction during the next cardiac cycle. The backed-up blood is gradually cleared over the subsequent few

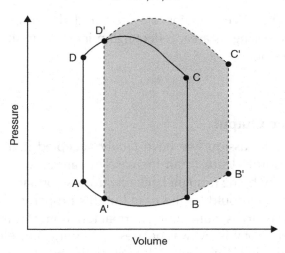

Fig. 62. Pressure volume loops illustrating the effects of increased preload (dashed lines).

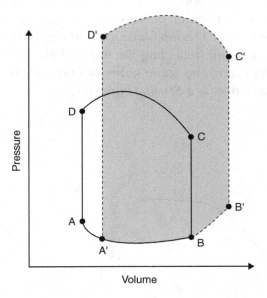

Fig. 63. Pressure volume loops illustrating the effect of increased afterload (dashed lines).

contractions but LVP remains elevated and the loop is stretched vertically, increasing its area, reflecting an increase in the work done per cardiac cycle.

5.4   Cardiac Output

In the above discussion, we have tacitly accepted that an increase in EDV necessarily leads to an increase in cardiac work, eventually resulting in EDV being brought back towards its normal value. It is not obvious that this should be the case but this property of the heart is vitally important to its maintaining normal function. This is the *Starling law.* The essence of this law was put best by Starling himself in his 1915 Linacre Lecture in Cambridge, "The law of the heart is thus the same as the law of muscular tissue generally, that the energy of contraction, however measured, is a function of the length of the muscle fibre". This is tantamount to saying that cardiac muscle usually operates on the steeply ascending portion of the tension-length curve, as we saw in Chapter 2. We can see this relationship more clearly by plotting some index of cardiac work (reflecting the force produced by the cardiac muscle) on the y-axis and some index of fibre length on the x-axis. This plot is referred to as a *Starling curve* (Fig. 64).

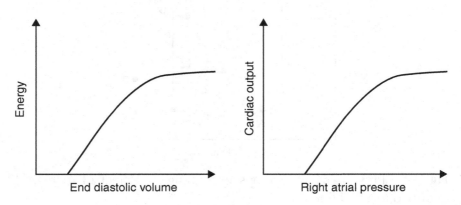

Fig. 64.   Starling curves.

We have established that the area enclosed within a pressure-volume loop gives the work done per cardiac cycle. If we make the rather sweeping assumption that ABP is constant then we can use change in volume, i.e. SV, as an index of cardiac work. Alternatively, if we also assume that HR is constant, we can plot CO on the y-axis. Similarly, we have established that atrial pressure can be used as an index of fibre stretch so we can plot RAP on the x-axis. This is the usual, and most useful, form of the Starling curve and is also called a *cardiac function curve*. Incidentally, the dotted line in Fig. 59 also reflects Starling's law: as EDV increases, cardiac work increases and LVP increases.

So far we have focused on the systemic circulation between the left ventricle and the right atrium. We must not forget that the heart also supports the pulmonary circulation, from the right ventricle to the left atrium. One of the more subtle effects of Starling's law is to balance the outputs of the right and left ventricles. This is important because, except transiently for a few beats, right ventricular output must equal left ventricular output. Even with a 1% imbalance between left and right ventricular outputs the pulmonary blood volume would increase by 350% within 30 minutes. As we will see later, this would lead to pulmonary congestion and oedema. Additionally, there are several common situations where transient imbalances between right and left cardiac outputs occur. For example, standing up results in the right ventricular output falling below the left ventricular output for a few beats. Furthermore, inspiration results in the right ventricular output rising above the left ventricular output for a few beats, as we shall again see later.

Starling's law also facilitates adaptation to alterations in heart rate. When heart rate decreases, diastole comes to occupy an increased proportion of the cardiac cycle. This results in greater ventricular filling and consequently increases in myocardial fibre length and thus stroke volume. A reduction in heart rate can then be fully compensated by the increase in stroke volume, such that the cardiac output can remain relatively constant.

In this way, Starling's law is the most important example of an *intrinsic* mechanism controlling CO. However, nervous and hormonal

LaPlace's Law and the Hook

In his original paper, Starling reported an interesting observation: at high degrees of stretch cardiac work actually began to *decrease*, giving a hook-like appearance. This is usually dismissed as an experimental artifact due to thrombosis, ischaemia or valvular regurgitation. However, there is in fact a theoretical explanation which is highly relevant in the failing, swollen heart. Modeling a ventricle as a sphere, we saw in Chapter 4 that LaPlace's law states:

$$P_i - P_o = \frac{2T}{r}$$

where P_i = ventricular pressure (Pa **or** mmHg)
        P_o = intra-thoracic pressure (Pa **or** mmHg)
        T = transmural tension (N·m^{-1})
        r = ventricular radius (m)

This is dealt with further in Appendix. We will assume that the intrathoracic pressure is zero. Thus, increasing the radius (volume) of a ventricle increases the tension in the muscle fibres required to produce a unit change in pressure, i.e. to generate pressure. This means that the degree to which the ventricular wall is curved determines how readily tension in the wall is converted into pressure.

This effect works directly against the Starling mechanism: as stretch increases, it becomes more difficult for the heart to contract. In the healthy heart at normal physiological RAPs, Starling's law easily wins out. However, in the unhealthy heart at high RAPs, LaPlace's law becomes increasingly important. Presumably the hearts used in Starling's experiments were not in the very best of health, explaining this phenomenon. This also explains why medical interventions that decrease cardiac filling are useful in patients with failing hearts.

factors also exert important *extrinsic* influences on CO. In the resting animal, basal activity in the parasympathetic nerves supplying the sinoatrial node exert a braking effect on HR. When an increase in CO is needed, this tonic inhibition is released and activity in the excitatory sympathetic nerves is increased. Both these changes result in an increase in HR, described as a positive *chronotropic* effect. In contrast, increased activity in the parasympathetic supply to the sinoatrial rode exerts a negative chronotropic effect. Unlike the parasympathetic nerves, the sympathetic nerves also innervate the ventricles. Activity in the sympathetic nerves also acts to increase ventricular contractility (energy developed at a particular degree of fibre stretch), described as a positive *inotropic* effect. In contrast, contractility may fall if the myocardium becomes diseased, described as a negative inotropic effect. Intrinsic regulation therefore describes changes in RAP resulting in movement along a single cardiac function curve while extrinsic regulation describes changes in myocardial contractility resulting in shifting between families of cardiac function curves (Fig. 65).

This all sounds good so far, but if the heart is replaced by a pump with a far higher capacity than the heart, CO is increased by only 5–10%. This is puzzling given that CO can increase by up to 600% during extreme exercise. Surprising as it seems, the heart is not the

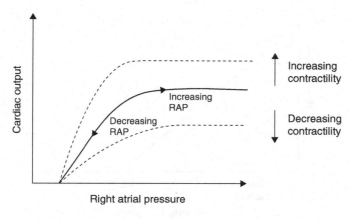

Fig. 65. "Families" of Starling curves.

key factor in determining the CO. To understand why, we need to now think about the blood returning to the heart, the VR.

5.5   Venous Return

Although it seems obvious, the heart cannot pump out more blood per unit time than it receives. This is tantamount to saying that CO is constrained by VR. As we know, VR is determined by the pressure gradient driving flow from the tissues to the right atrium:

$$MSFP - RAP = VR \cdot RVR$$

It is useful to plot the relationship between VR and RAP, referred to as a *vascular function curve* (Fig. 66).

As can be seen, this curve is a straight line for most of the range plotted. Rearranging this equation we have:

$$VR = \frac{MSFP}{RVR} - \frac{RAP}{RVR}$$

and comparing this with $y = mx + c$ we have $m = -\frac{1}{RVR}$. This gives the definition of RVR.

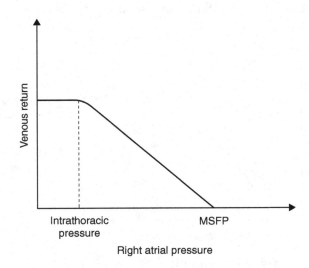

Fig. 66. Vascular function curve.

As RAP is increased, the difference between MSFP and RAP decreases, i.e. the gradient for flow decreases, and VR decreases: VR falls to zero when RAP = MSFP (around 7 mmHg). As RAP is made negative the curve diverges from the straight line predicted by this relationship and the definition of RVR ceases to apply. This occurs because, unlike arteries, veins are thin walled vessels and only remain open when a transmural pressure gradient forces them open. Considering the key vein returning blood to the right atrium (the *vena cava*), this pressure gradient is approximately given by the difference between the RAP and the intrathoracic pressure. Ordinarily RAP is more positive than intrathoracic pressure and the vein is forced open but as RAP becomes negative it tends to collapse, limiting VR. Notably the critical pressure at which collapse occurs will vary during the respiratory cycle as intrathoracic pressure varies. On inspiration, this critical pressure becomes more negative, increasing VR. This explains why replacing the heart with a high capacity pump has little effect on flow around the system: to increase CO, VR must increase and the only option open to the heart to increase VR is to suck by reducing RAP. Not only would this cause a right-shift on the cardiac function curve, but it might also not produce a meaningful increase in VR as the great veins soon collapse.

A more effective strategy for increasing VR is to increase MSFP. This will increase the pressure difference between the tissues and the RA and also shift the curve to the right along the x-axis. Notably, it will have no effect on the critical point at which venous collapse begins. We said earlier that MSFP is the pressure in the system at rest and as such is principally determined by the volume of blood in the circulation. A healthy male subject weighing 70 kg will have around 5 L of blood. The blood vessels have the capacity to accommodate around 80% of this volume without stretching. However, the final 20% does cause stretch and therefore contributes to the MSFP (Fig. 67). Most of the extra volume is accommodated in the veins rather than the arteries as they are so much more compliant.

The body can therefore achieve such an increase in MSFP by using the sympathetic nervous system to constrict veins surrounding the spleen, stomach and liver (the *splanchnic circulation*) and squeeze out

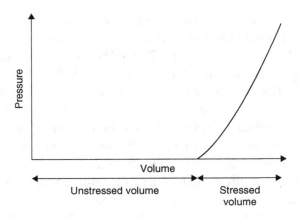

Fig. 67. "Stressed" and "unstressed" volumes.

its contents. By these mechanisms the useful circulating volume can be increased by up to 20%, significantly increasing MSFP. Of course, an intravenous infusion which increases circulating volume would have the same effect. In contrast, loss of blood (*haemorrhage*) would have the opposite effect. Just as for cardiac function curves, intrinsic regulation within the heart can produce changes in RAP which result in movement along a single vascular function curve while extrinsic regulation describes changes in MSFP resulting in shifting between families of vascular function curves (Fig. 68).

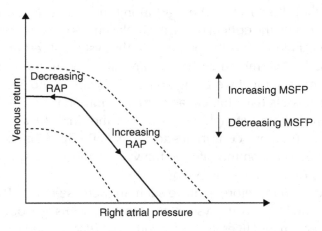

Fig. 68. "Families" of vascular function curves.

It is theoretically possible to affect CO by altering the value of RVR. This could be achieved by changing the diameter of the great veins using the sympathetic nervous system (*venoconstriction* and *venodilatation*). However, the resistance of the veins is in general negligible and therefore this has little effect on RVR. Note that this is a completely separate effect from autotransfusion. This said, important changes in RVR can take place and it is here that we must consider what is really meant by RVR in a little more detail. Defining the pressure gradient for VR as the difference between the pressure at the tissues (MSFP) and RAP is somewhat troubling since the tissues are not at one anatomical location, but rather are distributed throughout the body. In reality, the venous pressure is *topped-up* as blood flows back towards the heart: this can be modelled as a reduction in RVR. As skeletal muscle contracts and relaxes, it compresses veins running through it, squeezing blood back towards the heart. The back-flow of blood is prevented by venous valves. This *muscle pump* effect is particularly important in preventing blood from pooling in the legs while standing and is key to increasing flow during exercise. As already mentioned intrathoracic pressure falls during inspiration, sucking the great veins open and increasing VR. This effect is enhanced by the diaphragm descending during inspiration, compressing the abdominal veins. These effects are referred to as the *respiratory pump*. RVR does not therefore really represent a true resistance after all, it is really a made-up variable more related to local pressure gradients established by the muscle and respiratory pumps.

5.6   The Circulation as a Whole

While cardiac and vascular curves illustrate important properties, only one point on them really matters: the equilibrium point at which they meet. We have already seen that the heart cannot pump out more blood than it receives and therefore that CO cannot exceed VR. It could also be said that the heart must pump out all the blood it receives and therefore that VR cannot exceed CO.

This can be easily understood by plotting cardiac and vascular function curves on the same axis, in this case plotting flow around the entire system on the y-axis (Fig. 69).

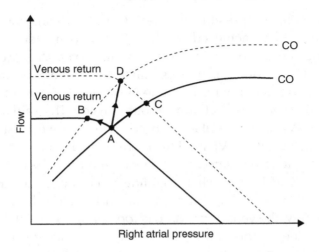

A ⟶ B : Increasing contractility without increasing
        MSFP decreases RAP but does not alter flow.

A ⟶ C : Increasing MSFP without increasing contractility
        increases flow but also increases RAP
        (seen in the failing heart).

A ⟶ D : Increasing both contractility and MSFP
        increases flow without increasing RAP.

Fig. 69.   Combined cardiac and vascular function curves.

In the circulation at rest, the two curves meet at point **A**, giving the resting RAP and flow. We will now consider the effect of increasing myocardial contractility and MSFP first separately and together. If contractility is increased without an increase in MSFP, for example following haemorrhage, the cardiac function curve shifts upwards but the vascular function curve remains unchanged. The result is that the equilibrium point shifts from **A** to **B**, i.e. RAP decreases but flow is not affected. Conversely, if MSFP is increased without a change in contractility, the vascular function curve shifts upwards but the cardiac function curve remains unchanged. This may be seen clinically in cardiac failure where various mechanisms act to expand the circulating volume. The equilibrium point shifts from **A** to **C**, i.e. flow increases at the expense of a large increase in RAP, confirming that the venous system can alter flow even when acting alone. However, if both contractility and MSFP are increased, the equilibrium point shifts from **A** to **D**: flow significantly increases while RAP remains unchanged. Note that we have not considered changes in RVR. These is dealt with in relation to exercise in Chapter 7. This is the ideal situation and is seen in the healthy circulation. We can therefore think of the vascular system being charged with the role of controlling flow, while the heart is concerned with controlling RAP.

5.7 The Pulmonary Circulation

The concepts discussed above on the whole apply equally well to the pulmonary circulation as to the systemic circulations. The main difference is that the pulmonary circulation operates at far lower pressures than the systemic circulation, with a mean pressure of just 13 mmHg in the pulmonary artery. While it is often said that the pulmonary circulation is able to operate at such low pressures because pulmonary vascular resistance is low, in reality this is only partly true.

To see this, we need to return to our discussion on flow. As described in the Appendix, flow will always occur from regions of high energy to regions of low energy. So far we have taken the rather simplified view that the pressure exerted by a fluid is the sole determinant of its energy per unit volume. However, two other factors make a contribution to the total energy of a fluid per unit volume. These are expressed in the *Bernoulli equation* for an incompressible fluid:

$$P + \rho g h + \frac{1}{2}\rho v^2 = \text{total energy per unit volume}$$

where ρ = density $(kg \cdot m^{-3})$
g = acceleration due to gravity $(m \cdot s^{-2})$
v = velocity $(m \cdot s^{-1})$
P = pressure (Pa **or** mmHg)
h = height (m)

and *total energy per unit volume* is measured in $J \cdot m^{-3}$ (**or** $J \cdot L^{-1}$).

Multiplying through by volume, we get:

$$PV + mgh + \frac{1}{2}mv^2 = \text{total energy}$$

where V = volume (m^3 or L), and *total energy* is measure in J.

Thus, for a full description of the energy of a fluid and of energy gradients, we must also consider contributions made by the gravitational potential energy and kinetic energy of the fluid. In both the systemic and pulmonary circulations, gravitational potential energy differences are negligible and this term can be ignored. In the systemic circulation, the contribution made by kinetic energy is negligible compared with that made by pressure and this term can be ignored in our simplified treatment. However, in the pulmonary circulation, the contribution made by kinetic energy is significant. This leads to a view of the left ventricle as a pressure generator and the thinner-walled right ventricle as a flow generator.

The most important consequence of the low pressures at which the pulmonary circulation operates is that it is rendered particularly sensitive to the effects of increased pressures. This can result either from pressure backing-up from a failing left ventricle or from increased pulmonary vascular resistance. As discussed in Chapter 4, local increases in pulmonary vascular resistance are an important mechanism for the redirection of blood away from poorly-ventilated regions of the lung to better-ventilated regions (*hypoxic pulmonary vasoconstriction*). However, when the lung experiences generalised hypoxia, the result may be a global increase in resistance and pulmonary arterial hypertension. This is a major problem: it will lead to oedema formation by the *Starling filtration-reabsorption mechanism*. This may result in a significant decrease in the rate at which gas diffuses between the alveoli and the pulmonary capillaries, as explained in relation to the *Fick equation* in Chapter 4.

5.8 The Starling Filtration-Reabsorption Mechanism

So far we have applied Darcy's law to longitudinal flow along blood vessels occurring as a result of pressure differences along their lengths. Darcy's law can also be applied to the flow of fluid across a vessel wall. In this setting, ΔP represents the total pressure difference across the vascular wall between the intravascular compartment and the interstitial compartment. This time, however, we not only need to consider hydrostatic pressure difference but also *osmotic pressure difference* (see Chapter 6, Further Thoughts). An osmotic pressure difference arises because while the plasma has a high protein content, the interstitial fluid contains virtually none. Thus while the hydrostatic pressure inside the vessel is greater than the hydrostatic pressure outside the vessel, the osmotic pressure inside the vessel is less than the osmotic pressure outside the vessel (Fig. 70).

We can therefore apply Darcy's law and write:

$$(P_i - P_o) - (\pi_i - \pi_o) = \dot{V} \cdot R$$

where P_i = hydrostatic pressure inside the vessel (Pa **or** mmHg)
P_o = hydrostatic pressure outside the vessel (Pa **or** mmHg)
π_i = colloid osmotic pressure inside the vessel (Pa **or** mmHg)
π_o = colloid osmotic pressure outside the vessel (Pa **or** mmHg).

This equation assumes that a capillary membrane acts as a perfect semipermeable membrane, completely excluding plasma proteins. Of course, as with everything else, in reality it is imperfect and some protein leaks across and a smaller-than-expected osmotic pressure acts. This problem can be circumvented by introducing a dimensionless *reflection coefficient*, σ. Here $\sigma = 1$ means that the membrane is completely impermeable to plasma proteins and $\sigma = 0$

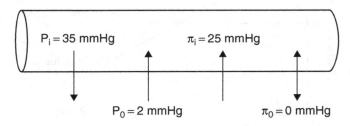

Fig. 70. Starling forces across capillary membranes.

means that it is completely permeable to plasma proteins. Thus:

$$(P_i - P_o) - \sigma(\pi_i - \pi_o) = \dot{V} \cdot R$$

where σ = reflection coefficient for plasma proteins (dimensionless).

In this context, it is conventional to refer to conductance, or *filtration coefficient*, rather than resistance. This is given by the product of membrane permeability and surface area. We can therefore rewrite the above equation as:

$$\dot{V} = K_f \cdot [(P_i - P_o) - \sigma(\pi_i - \pi_o)]$$

where K_f = filtration coefficient (representing conductance in Darcy's law, $m^4 \cdot s^{-1} \cdot kg^{-1}$ **or** $L \cdot min^{-1} \cdot mmHg^{-1}$).

This is the *Starling filtration-reabsorption equation* and is of fundamental importance in many areas of physiology. Normally P_i varies between around 35 mmHg proximally in capillaries to around -15 mmHg in venules, P_o is constant at -2 mmHg, π_i is 25 mmHg, π_o is zero and σ is around 0.9. Putting these numbers into the equation gives a net gradient for fluid movement from the vessels into the tissues (positive $\dot{V}$) in the capillaries and a net gradient for fluid movement from the tissues into the vessels (negative $\dot{V}$) in the venules. This gives the solid line in Fig. 71.

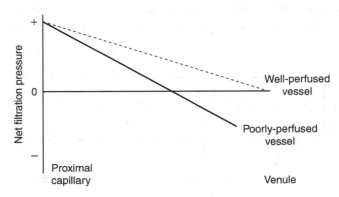

Fig. 71. Filtration of fluid across capillary membranes.

However, this does not turn out to be correct *in vivo*. Instead, in well perfused vessels the net pressure gradient for filtration remains positive even in the venules, giving the dashed line. This is probably due to increases in π_o along the capillary as protein escapes.

As a result, mechanisms must be in place to prevent the build-up of excess fluid in the tissues (*oedema*). P_i is largely controlled by precapillary sphincters. Contraction of these sphincters increases resistance and decreases intravascular hydrostatic pressure and cyclical contraction followed by relaxation results in transient net reabsorption. The *lymphatic system* is also of key importance in removing fluid. Naturally, if P_i decreases as a result of a decrease in ABP or MSFP, for example in haemorrhage, there will be a net gradient for reabsorption. This allows the interstitium to buffer the plasma volume. Furthermore, P_i may increase when ABP or MSFP increase, for example in heart failure, and lead to an increase in flow. Similarly, if plasma protein content decreases (*hypoproteinaemia*), for example because of protein loss in renal disease or in starvation, π_i will decrease, leading to an increase in flow. K_f can also increase, whether due to an increase in membrane permeability as a result of damage in hypoxia or an increase in surface area (only really important in the kidney), leading to an increase in flow (see Chapter 6). This will result in oedema unless it is compensated for by an increased rate of lymphatic drainage.

FURTHER THOUGHTS

5.9   Deviations from Darcy's Law

As noted earlier, Darcy's law is only truly valid if a number of rather stringent conditions apply. These conditions are not necessarily met in the cardiovascular system, making these equations only approximately true.

Firstly, Darcy's law applies only when flow is laminar, i.e. when fluid flows in parallel layers between which there are no interactions. Under ideal conditions, laminar flow is likely to occur when *Reynold's number* is less than a critical value.

$$N_R = \frac{vd\rho}{\eta}$$

where N_R = Reynold's number (m^{-2}·s^{-2})
  v = velocity (m·s^{-1})
  d = diameter of tube (m)
  ρ = density of fluid (kg·m^{-3})
  η = viscosity of fluid (kg·m^2·s^{-1}).

As Reynold's number increases, flow becomes increasingly turbulent, with swirls of fluid no longer moving parallel to the direction of flow but also in other directions and layers interacting with each other. When Reynold's number exceeds a critical value, flow is likely to be predominantly turbulent. In this situation, energy is wasted and a greater pressure drop occurs per unit length of tube, i.e. Darcy's law needs to be modified such that rather than being raised to the power one, the driving pressure gradient is raised to some power less than one. As can be seen, Reynold's number is proportional to vessel diameter and to velocity, so turbulent flow is likely to occur in large-diameter vessels where velocity is high. Turbulence almost always occurs in the aorta and pulmonary artery during systole and may also occur in other arteries when velocity increases during exercise. Incidentally, Reynold's number is inversely proportional to viscosity so turbulent

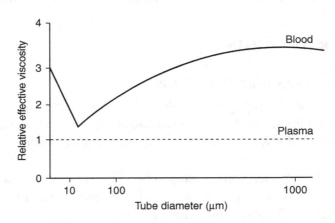

Fig. 72. Fåhraeus-Lindqvist effect.

flow may occur in conditions such as anaemia where viscosity is low. This sometimes results in an audible *murmur*.

Secondly, Darcy's law requires that the fluid flowing is homogenous and has a constant viscosity. In reality, blood is a suspension of cells in plasma. Friction forces between adjacent laminae of fluid move cells towards the central axis of the tube, effectively decreasing the viscosity of the fluid near the vessel wall (*marginal layer*) and increasing the viscosity of the fluid near the central axis (*axial layer*). This is referred to as *axial streaming*. It is most apparent at high flow velocities, meaning that effective viscosity decreases with increasing velocity. It follows that since flow through arterioles and capillaries is slow, effective viscosity should be high. Curiously, this is not the case. The effective viscosity of any suspension, including blood, is empirically found to decrease with decreasing tube diameter. This phenomenon begins at tube diameters of around 1 mm (small arteries) and continues until a tube diameter of around 5 μm (capillaries) is reached, where effective viscosity is as low as that of plasma. This is referred to as the *Fåhraeus-Lindqvist effect* (or *sigma effect*) (Fig. 72). Incidentally, effective viscosity increases again as tube diameter is further decreased.

The reasons for this phenomenon are incompletely understood and there are several possible explanations, as illustrated in Fig. 73. One,

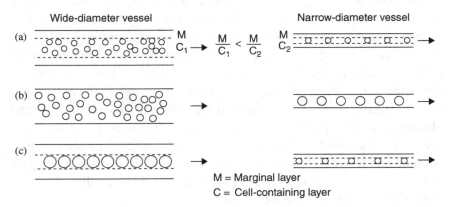

Fig. 73. Possible explanations for the Fåhraeus-Lindqvist effect.

illustrated in Fig. 73(a), is based on the fact that the thickness of the marginal layer is only slightly affected by the diameter of the tube. Therefore as tube diameter decreases, the marginal layer comprises an increasing proportion of the total vessel diameter. As we know, cells tend to move away from the marginal layer towards the axial layer and therefore the viscosity of the marginal layer is lower than that of the axial layer. As the proportion of the total vessel diameter taken up by the marginal layer increases, effective viscosity decreases. A second possible explanation (Fig. 73(b)) arises because capillaries have diameters close to those of red blood cells. Therefore red blood cells must line up in an orderly manner (*bolus flow*), reducing turbulence and thereby reducing effective viscosity. A third possibility (Fig. 73(c)) relates to flow profile. Thinking again of a flowing fluid as a series of laminae, the first lamina in contact with the vessel wall has a velocity of zero since there are strong friction forces between the lamina and the vessel wall (the *zero-slip condition*). The second lamina will slide slowly past the first lamina, i.e. has a velocity greater than zero. The third lamina will slide past the second lamina, i.e. has a velocity of greater than that of the second lamina. This continues as we move away from the wall of the vessel: a maximum velocity is reached at the in the axial stream, resulting in a parabolic velocity profile. This applies in large-diameter vessels just as it does in small-diameter vessels. As we know, cells tend to move towards the axial stream. The mean velocity of cells is therefore greater than the mean velocity of plasma. If flow is to remain constant as blood moves from a large-diameter vessel into a small-diameter vessel, then its velocity must increase. The velocity of cells will increase by the same proportion as the velocity of plasma. However, since the cells are already moving faster than the plasma the absolute increase in the velocity of cells will be greater than the absolute increase in the velocity of plasma. This must mean that at a given flow the number of cells per unit volume of blood is smaller in a small-diameter vessel than in a large-diameter vessel. This will mean that the effective viscosity of blood is smaller in a smaller-diameter vessel than in a large-diameter vessel.

Returning to Darcy's law, the third condition requires that the tube through which a fluid flow be rigid. This is tantamount to saying that changing the pressure inside the tube has no effect on radius and therefore on volume, and therefore that the compliance of the wall of the tube is zero. In fact, blood vessels are elastic to varying degrees and increases in pressure result in increases in radius and therefore in volume i.e. they have a compliance of greater than zero. As shown in the Appendix:

$$R_{fluid} \propto \frac{1}{r^4}$$

where r = radius of pipe (m).

An *increase* in intravascular pressure will therefore now result in a *decrease* in vascular resistance as a result of an increase in radius. Given that resistance now depends on intravascular pressure, flow will no longer be directly proportional to the driving pressure gradient. In fact, with any pressure gradient flow will be greater than predicted by Darcy's law (Fig. 74).

A further important consequence of vessel walls being non-rigid is that vessels can collapse inwards. We can imagine that tensile forces in stretched vessel walls will generate a force pulling inwards towards the centre of the tube, tending to cause collapse. A certain minimum

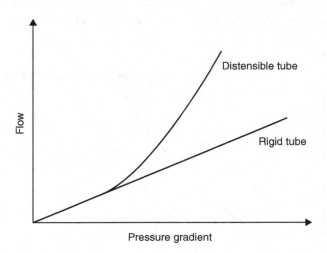

Fig. 74. Relationships between pressure gradient and flow in rigid and distensible tubes.

intravascular pressure is needed to exert a sufficient opposing force to prevent a vessel from collapsing. If the intravascular pressure is less than this *critical closing pressure* the vessel will collapse. As discussed in Derivations, for a cylindrical vessel:

$$P_{cc} = \frac{T}{r}$$

where P_{cc} = critical closing pressure (Pa **or** mmHg).

However, we cannot predict on a theoretical basis whether the concept of critical closing pressure is relevant physiologically: it might be that the critical closing pressure is less than MSFP. We would expect the critical closing pressure to take its largest value when transmural tension is large and radius is small, i.e. in constricted arterioles. Experiments demonstrate that the critical closing pressure is around 20 mmHg in such vessels: flow ceases when intravascular pressure is less than this value. This means that the curve relating flow and the driving pressure gradient does not pass through the origin as predicted by Darcy's law. This can be seen in Fig. 75.

Importantly if ABP drops as a result of cardiac failure, or if MSFP drops as a result of haemorrhage, intravascular pressure may fall below the critical closing pressure, resulting in vessel collapse. This may be exacerbated by sympathetically-mediated vasoconstriction, intended

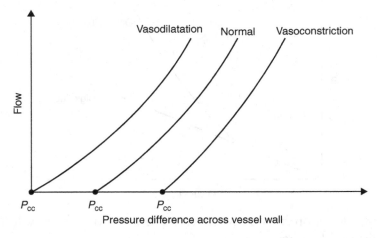

Fig. 75. Critical closing pressure.

to increase TPR and therefore ABP, paradoxically increasing the critical closing pressure.

5.10   Measuring Cardiac Output

CO is a hugely important but somewhat hard-to-measure quantity. CO can be thought of as the product of heart rate and stroke volume, the volume of blood ejected by each ventricle per contraction. It is easy to measure the heart rate by feeling the pulse rate in the radial artery. SV is properly estimated using echo-cardiography to visualise the heart. However, a rough estimate of the SV can be obtained more simply. This is achieved by assuming that the entry of 2 ml of blood into the aorta results in an increase in pressure of 1 mmHg. It follows that the stroke volume in millilitres is then equal to two times the *pulse pressure* (systolic pressure minus diastolic pressure). Clearly this relationship is only true if we assume a certain value for the vascular compliance (stretchiness): it applies only in young people.

More accurate values for the CO can be obtained in a number of ways. In experimental animals, CO has been measured using electro-magnetic or ultrasound flow probes implanted around the aorta. Information can then be recorded directly or transmitted to a receiving unit (telemetry), leaving the animal free to move. Notably faithful measurement of CO by these techniques requires that the probe fit tightly around the aorta. This has the potential to reduce the diameter of the aorta, increase its resistance and therefore by Darcy's law, give an erroneously low value (see Appendix). Nonetheless, much important information has been obtained using these sorts of recording techniques.

Despite their accuracy, flow-probes are not a useful technique for the clinical measurement of CO. Two straightforward and less invasive methods are available, the *Hamilton indicator dilution technique* and the *Fick principle*.

5.10.1   *The Hamilton technique*

In the Hamilton technique, a known mass of dye is rapidly injected into a central vein (Fig. 74). The dye mixes into the plasma, enters the heart and is pumped out. Peripheral arterial blood is continually sampled and the concentration of the dye in the blood is plotted against time (Fig. 76).

As the indicator arrives at the peripheral artery, the concentration recorded begins to increase. Concentration then reaches a peak, begins to decline exponentially (the left ventricle ejects a fixed proportion of its contents per contraction) and plateaus. This takes place as the first of the dye returns to the peripheral artery for a second time, having travelled all the way around the shortest closed loop in the cardiovascular system, the coronary circulation. Extrapolating the initial phase of decline back to the x-axis gives a complete profile of changes in indicator concentration during one complete pass around the cardiovascular system. The volume of blood which has passed through the artery during one complete pass is given by:

$$\text{volume} = \frac{\text{mass}}{[\text{indicator}]_{\text{mean}}}$$

where volume = blood volume (m^3 **or** L)

[indicator]$_{\text{mean}}$ = mean indicator concentration

mass = mass of indicator injected ($kg \cdot m^{-3}$ or kgL^{-1}).

Dividing this value, the time taken for one complete pass, gives the cardiac output of plasma. Happily we can simplify things: the shaded *area under the curve* gives the product of the mean concentration and the time. Thus:

$$CO_{\text{plasma}} = \frac{\text{mass}}{\text{AUC}}$$

where CO_{plasma} = cardiac output of plasma ($m^3 \cdot s^{-1}$ **or** $L \cdot min^{-1}$)

AUC = area under the curve ($kg \cdot s \cdot m^{-3}$ **or** $kg \cdot min \cdot L^{-1}$).

This method gives the cardiac output of *plasma*. However, the blood consists of cells (mostly red blood cells) suspended in plasma and these collectively have a significant volume. The proportion of the blood comprised of red cells is referred to as the *haematocrit, hct* (although incidentally *haematocrit* was the original term for the apparatus used to measure this quantity). This is easily measured. A blood sample is taken and its volume is determined. The sample is then spun in a centrifuge causing the red cells to settle to the bottom. The plasma is then poured off, its volume is measured and the *hct* is calculated. Now:

hct = proportion of blood represented by cells

and since blood is made up of cells and plasma only,

$(1 - hct)$ = proportion of blood represented by plasma

hence,

$$\text{volume of plasma} = \text{volume of blood} \cdot (1 - hct)$$

and

$$\text{volume of blood} = \frac{\text{volume of plasma}}{(1 - hct)}$$

so

$$CO = \frac{CO_{plasma}}{1 - hct}$$

Modern *thermal dilution* techniques use cold saline instead of a dye. Not only does this avoid potential toxicity, but it also largely eliminates *second-pass effects* as the saline has warmed to body temperature by the time it has passed through the coronary circulation.

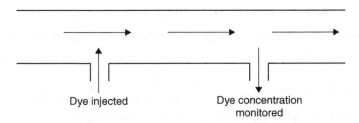

Dye injected   Dye concentration
                                monitored

Fig. 76. The Hamilton technique.

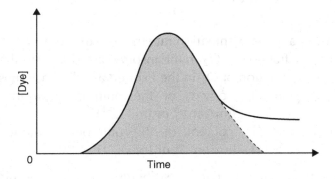

Fig. 77. Results obtained using the Hamilton technique.

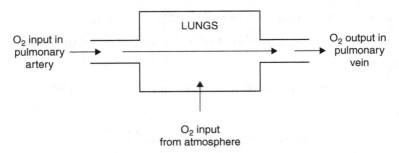

Fig. 78. The Fick principle.

5.10.2 *The Fick principle*

The Fick Principle is nothing more that the Law of the Conservation of Mass. In its original formulation, it simply states that, at steady state, the rate at which O_2 arrives in the alveolar capillary blood equals the rate at which it leaves, i.e. *input = output* (Fig. 76).

It can already be seen from the diagram that

O_2 extraction = CO · arterio − venous concentration difference

and hence

$$CO = \frac{O_2 \text{ extraction}}{\text{arterio} - \text{venous concentration difference}}$$

This is the Fick principle. It can perhaps more usefully be written as:

$$CO = \frac{\dot{V}_A \cdot (F_{IO_2} - F_{EO_2})}{C_{O_2\,PV} - C_{O_2\,PA}}$$

where V_A = alveolar ventilation rate ($m^3 \cdot s^{-1}$ **or** $L \cdot min^{-1}$)
  F_{IO_2} = fraction of O_2 in the inspired air (dimensionless)
  F_{EO_2} = fraction of O_2 in the expired air (dimensionless)
  $C_{O_2\,PA}$ = O_2 content of the pulmonary arterial blood ($(m^3) \cdot m^{-3}$ **or** $(ml) \cdot L^{-1}$)
  $C_{O_2\,PV}$ = O_2 content of the pulmonary venous blood ($(m^3) \cdot m^{-3}$ **or** $(ml) \cdot L^{-1}$).

Of these parameters, V_A can be easily measured using a spirometer. F_{IO_2} and F_{EO_2} are also easily measured by taking atmospheric and

Derivation of the Fick Equation

The input has two components, from the atmosphere as a result of alveolar ventilation and from the blood vessel supplying the lungs, the pulmonary artery. The quantity of O_2 entering the alveolar capillary blood per unit time due to alveolar ventilation is given by

$$\dot{V}_A \cdot (F_{IO_2} - F_{EO_2})$$

Since the blood flow through the pulmonary artery is equal to the cardiac output then:

rate of O_2 entry into alveolar capillary blood $= C_{O_2PA} \cdot CO$

here $C_{O_2PA} = O_2$ content of the pulmonary arterial blood $((m^3) \cdot m^{-3}$ **or** $(ml) \cdot L^{-1})$.

So:

$$\text{input} = \dot{V}_A \cdot (F_{IO_2} - F_{EO_2}) + C_{O_2PA} \cdot CO$$

The output has just one component, O_2 carried away in the pulmonary vein. Blood flow in the pulmonary vein is also equal to the cardiac output. The output is therefore the quantity of O_2 leaving the alveolar capillary blood per unit time from the pulmonary vein:

$$\text{output} = C_{O_2PV} \cdot CO$$

where $C_{O_2PV} = O_2$ content of the pulmonary venous blood $((m^3) \cdot m^{-3}$ **or** $(ml) \cdot L^{-1})$.

Now input = output so:

$$\dot{V}_A \cdot (F_{IO_2} - F_{EO_2}) + C_{O_2PA} \cdot CO = C_{O_2PV} \cdot CO$$

which rearranges to give:

$$CO = \frac{\dot{V}_A \cdot (F_{IO_2} - F_{EO_2})}{C_{O_2PV} - C_{O_2PA}}$$

expired gas samples respectively and passing them through a gas analyser. However, measuring $C_{O_2 PA}$ and $C_{O_2 PV}$ is somewhat more of a problem. $C_{O_2 PA}$ must be measured from a sample of mixed venous blood obtained from the right ventricle or the pulmonary artery. In a rather heroic experiment in 1929, Forssman passed a urinary catheter into his right ventricle via a vein in his arm, allowing pulmonary arterial blood to be sampled for the first time. This permitted the first application of the technique Fick had proposed almost sixty years earlier. Mixed venous blood samples are obtained in much the same way nowadays. Notably, the O_2 content of peripheral venous blood is highly variable and does not usefully represent $C_{O_2 PA}$. Happily the same is not true of peripheral arterial blood: the O_2 content of blood obtained from the brachial or femoral arteries closely approximates $C_{O_2 PV}$. Thus:

$$CO = \frac{O_2 \text{ extraction}}{\text{arterio} - \text{venous concentration difference}}$$

CHAPTER 6

The Kidney and Body Fluids

The maintenance of a constant internal environment is the very essence of homeostasis. The kidney, the organ with the primary responsibility for regulation of the volume and composition of body fluids, takes a key role in this process. The structure of the kidney's functional unit, the *nephron*, is shown in Fig. 79. This comprises three key parts, dealt with in detail later: the *glomerulus*, the *tubule* and the *collecting duct*. The vital importance of the kidney in these matters is clear from the consequences of renal failure. In this situation, toxins and ionic salts accumulate in the body. Accumulation of Na^+ results in the retention of water to maintain *osmolarity* (see Further Thoughts) and this leads to expansion of extracellular fluid volume. Not only will this lead to oedema, but expansion of the blood volume leads to an increase in MSFP that can in turn result in heart failure (see Chapter 4). If plasma osmolarity is also compromised, then the situation becomes even worse. Increased plasma osmolarity results in an osmotic drive for water to leave cells. Conversely, decreased plasma osmolarity results in a drive for water to enter cells. The most malign consequence of this is that the brain swells. Since the skull (*cranium*) is a closed compartment, this increases the intracranial pressure and may compromise the blood flow to the brain. Of the other substances that accumulate, K^+ poses the greatest threat to homeostasis. An excess of K^+ in the ECF results in the depolarisation of cardiac muscle cells, ultimately leading to death from cardiac arrhythmia (see Chapter 1).

In this chapter, we describe the main principles by which the kidney regulates the volume and composition of the body fluids. At its simplest level, this is achieved by regulating output in the face of a variable

145

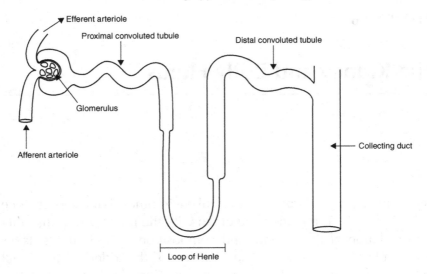

Fig. 79. The nephron.

input from metabolism and from the gut (see Chapter 3). It does this by selectively regulating the removal of specific substances from the plasma into urine. Thus, by varying the rate of production and composition of the urine, the kidney fulfills the main part of its function.

6.1 The Formation of Urine

The fluid that will become urine is initially formed in the glomeruli by *filtration*. It is then modified as it passes along the tubules by selective *reabsorption* and *secretion*. Here we will consider each of these processes in turn, before considering illustrative examples of how the kidney handles particular substances.

6.1.1 Filtration

Filtration is the first step in the formation of urine. In this process, blood plasma is forced across a semi-permeable membrane (*filtration barrier*) to form *glomerular filtrate*. This filtration barrier has three key components, all of which carry a negative charge: the glomerular capillary wall, support cells and the tubular membrane. The filtration barrier prevents particulate components of plasma with molecular weights of greater than 70 kDa from crossing into the glomerular filtrate. Notably, haemoglobin has a molecular weight of around 69.8 kDa: were it not sequestered in red blood cells it would be filtered at the glomerulus and might be lost in the urine. As noted in Chapter 1, such protein molecules are usually negatively charged at a physiological pH. The negative charge on the filtration barrier therefore repels these particles. This results in their being filtered to a lesser degree than positively charged particles of similar sizes. The glomerular capillary blood is therefore slightly negatively charged with respect to the tubular fluid. The resulting electrical energy difference (see Chapter 1) creates an additional drive for the filtration of small inorganic ions. Ultimately, this means that negatively charged ions are filtered to a lesser degree than positively charged ions of similar sizes. Therefore, while the composition of this *primary filtrate* is unregulated, it does differ somewhat from that of plasma. Nevertheless, it has the same osmolarity as plasma (it is *isosmotic* with plasma).

6.1.2   *Glomerular filtration rate*

Just as we saw in Chapter 3 that it was important to control flow through the digestive tract, it is also important to control flow through the renal tubules. On the one hand, a large increase in this flow would result in there being insufficient time to modify the primary filtrate before it leaves the tubules. On the other hand, a large decrease would prevent the kidney from excreting toxins at a sufficient rate to prevent accumulation in the body. Thus, control over the rate of filtration, the *glomerular filtration rate* (GFR), is critical to the function of the kidney and hence to homeostasis. In young, healthy adults GFR is maintained at around $125 \, \text{ml} \cdot \text{min}^{-1}$.

Clearly all of the fluid that is filtered at the glomerulus must have first arrived in the glomerular capillary blood. It therefore follows that a normal GFR in turn requires a normal longitudinal flow of blood along the glomerular capillaries (*renal blood flow*, RBF).

By Darcy's law,

$$\Delta P = \dot{V} \cdot R_{\text{fluid}}$$

where ΔP = pressure difference (Pa **or** mmHg)

$\dot{V}$ = flow (m^3·s^{-1} **or** L·min^{-1})

R_{fluid} = resistance to fluid flow (kg · s^{-1} · m^{-4} **or** mmHg·L^{-1}·min)

It follows that RBF should vary with ABP. However, rather remarkably, RBF is maintained almost constant over the full physiological range of mean ABPs, between 95 and 165 mmHg. This is achieved by altering the resistance of the blood vessel supplying the glomerular capillaries, the *afferent arteriole*. The process by which a constant afferent arteriolar blood flow is maintained is described as *renal autoregulation*. Renal autoregulation has two components, the *myogenic mechanism* and the *tubuloglomerular feedback* mechanism.

The myogenic mechanism is by far the simplest and can be demonstrated even in isolated segments of afferent arteriole. Modelling the afferent arteriole as a thin walled cylinder, LaPlace's law (see Appendix) states that:

$$P_i - P_o = \frac{T}{r}$$

where P_i = pressure inside the vessel (Pa **or** mmHg)

P_o = pressure outside the vessel (Pa **or** mmHg)

T = transmural tension in the wall per unit length (N·m^{-1} **or** mmHg·m^{-1})

r = radius (m).

If afferent arteriolar pressure increases, then the stretch experienced by the vessel wall increases. This increased stretch opens stretch activated ion channels in the afferent arteriolar wall. This leads to the depolarisation of smooth muscle cells in the vessel wall and then to muscle contraction. This in turn decreases the radius of the afferent arteriole. From Poiseuille's law (see Appendix):

$$R_{fluid} \propto \frac{1}{r^4}$$

where r = radius of pipe (m).

Decreasing the radius therefore increases resistance. This returns RBF, and therefore GFR, to normal. The decrease in radius also decreases the stretch experienced by the vessel wall, again by LaPlace's law. This closes the stretch activated ion channels in the afferent arteriolar wall and completes the negative feedback cycle.

Tubuloglomerular feedback is somewhat less straightforward and is still the subject of experimental study. This mechanism involves the *macula densa*, a region of modified muscle cells forming part of the thick ascending limb of the *loop of Henle*, in co-operation with the afferent arteriole. The macula densa monitors the flux of Cl^- (probably using the $Na^+K^+2Cl^-$ carrier discussed in detail later), and in that way monitors the GFR. If GFR changes, the macula densa alters the production of a messenger that diffuses back to the afferent arteriole and affects its resistance, thereby altering RBF and normalising GFR. There is still much debate as to the identity of this messenger, and indeed as to whether it is a vasoconstrictor factor produced in response to an increase in GFR, or a vasodilator produced in response to a decrease in GFR. The current prevailing thought is that it is a vasoconstrictor and is in fact ATP. This explains the earlier observation that macula densa cells contain far more abundant mitochondria than they could possibly need for their own energy consumption. Ordinarily, the abundance of mitochondria in a cell is well fitted to its energy need.

Having established how a constant and sufficient RBF is maintained, we are now in a position to discuss the mechanism of filtration in more detail. Again applying Darcy's law, the total transverse flow of fluid across the filtration barrier (GFR) in the kidney must depend on both the resistance to flow and the driving pressure gradient. In this case, it is conventional to talk in terms of the conductance, rather than the resistance, of the filtration barrier (the *filtration coefficient*). The GFR is then determined by the *Starling filtration-reabsorption* equation. Notably, this same equation defines the transverse flow of fluid across capillary membranes everywhere in the body, not just in the kidney. Hence these same factors are also key in the formation of interstitial fluid. This was described in Chapter 5. We can now discuss each relevant factor in turn.

$$\text{GFR} = K_f \cdot [(P_i - P_o) - \sigma(\pi_i - \pi_o)]$$

where K_f = filtration coefficient (representing conductance in Darcy's
        law, $m^4 \cdot s^{-1} \cdot kg^{-1}$ **or** $L \cdot min^{-1} \cdot mmHg^{-1}$)
P_i = hydrostatic pressure inside the vessel (Pa **or** mmHg)
P_o = hydrostatic pressure outside the vessel (Pa **or** mmHg)
σ = reflection coefficient for plasma proteins (dimensionless)
π_i = colloid osmotic pressure inside the vessel (Pa **or** mmHg)
π_o = colloid osmotic pressure outside the vessel (Pa **or** mmHg).

Intuitively, the filtration coefficient must depend on two factors: the
size of the holes in the filtration barrier and the surface area over which
filtration can occur. First, the size of the holes in the filtration barrier
(*hydraulic permeability*) determines the resistance of each possible
path through which fluid could flow by Poiseuille's law. In *acute
renal failure*, damage to the filtration barrier may increase the sizes
of these holes, increasing GFR in an uncontrolled manner. Secondly,
the surface area over which filtration can occur determines the number
of possible paths: the greater the number of possible paths arranged
in parallel, the lower the total resistance (see Appendix). This surface
area can be altered physiologically by contraction or relaxation of the
renal mesangial cells. ATP released as part of the tubuloglomerular
feedback mechanism in response to increased GFR causes contraction
of these cells, decreasing GFR.

Changes in this pressure gradient driving transverse flow produce
proportionate changes in GFR. This pressure gradient has both
hydrostatic and osmotic components.

The hydrostatic pressure inside the glomerular capillaries is ordinarily high (around 60 mmHg at the beginning of the glomerular
capillaries), while that inside the tubules is low (around 15 mmHg).
The resulting hydrostatic pressure difference of circa 45 mmHg drives
the movement of fluid from the glomerular capillaries into the tubules.
The hydrostatic pressure inside the glomerular capillaries is determined
by the afferent arteriolar pressure. This is in turn dependent on
ABP. Fortunately, the mechanisms in place to ensure a constant and
sufficient RBF also regulate afferent arteriolar pressure. An increase
in ABP would, by renal autoregulation, result in increased afferent
arteriolar resistance. Again applying Darcy's law, this would increase

the drop in pressure per unit length of the afferent arteriole and thereby normalise afferent arteriolar pressure. A decrease in ABP would have the converse effect on autoregulatory mechanisms. However, it is important to note that large decreases in ABP may exceed the regulatory capability of these mechanisms, compromising GFR.

As we have established, the filtration barrier is freely permeable to solutes in plasma. Therefore, such solutes do not generate an osmotic pressure difference across the filtration barrier (they constitute ineffective osmolytes, see Further Thoughts) despite being the predominant contributors to plasma osmolarity. However, the filtration barrier is impermeable to the protein components of plasma. This is reflected in the reflection coefficient being approximately equal to 1. This results in protein being present in the glomerular capillary blood but absent from the tubular fluid. Given that the osmolarity contributed by plasma proteins in the glomerular capillary blood is around 1 mOsM and that standard body temperature is 310 K, the *colloid osmotic pressure* can be calculated using van't Hoff's law (see Further Thoughts):

$$\pi_i = CRT$$
$$\pi_i \simeq 1 \cdot 8.31 \cdot 310 \simeq 2600 \, \text{N} \cdot \text{m}^{-2}$$
$$\pi_i \simeq 19 \, \text{mmHg}$$

In contrast, π_o must be effectively 0 mmHg. The result is a colloid osmotic pressure difference of around 19 mmHg across the filtration barrier that tends to drive water from the tubules into the glomerular capillaries. This colloid osmotic drive is small in comparison to the hydrostatic drive. The colloid osmotic pressure inside the glomerular capillaries may increase in clinical disorders where plasma protein content increases, such as *amyloidosis*. It may decrease in conditions where plasma protein content decreases, such as during starvation or in liver failure (this highlights the vital role of the liver in the synthesis of plasma proteins). This will also occur as a result of the intravenous infusion of solutions that do not contain protein.

6.1.3 *Reabsorption*

Reabsorption refers to the movement of substances from the tubular fluid back into the blood (after they have been filtered). Since ultrafiltration is a non-selective process, selective reabsorption provides a means by which the plasma concentration of substances can be regulated. If the body always needs to retain a substance, then reabsorption must take place at the same rate as filtration. Reabsorption will then be complete and none of the substance will be lost in the urine. In contrast, if the kidney is to be used to regulate the amount of a substance in the body, then the rate of reabsorption must be regulated independently of the rate of filtration. As we will see later, glucose and Na^+ respectively provide examples of such substances.

In both cases, most reabsorption takes place in the *proximal convoluted tubule* (see Fig. 77). Indeed, under normal conditions, around 67% of the filtered Na^+, Cl^- and water, as well as 85% of filtered HCO_3^- and all the filtered glucose and amino acids, are reabsorbed. However, this reabsorption is largely unregulated and, as we will see in greater detail later, the regulation of reabsorption occurs mostly in the distal tubules.

6.1.4 *Secretion*

Secretion is the opposite process to reabsorption and refers to movement of substances from the blood into the tubular fluid. This allows substances to be removed from the body more effectively than by filtration alone. As with reabsorption, altering the rate of secretion of a substance also enables regulation of the plasma concentrations.

6.2 Examples of Renal Handling of Substances

We can illustrate these principles by considering the renal handling of three substances: inulin, glucose and para-aminohippuric acid (PAH).

6.2.1 *Inulin*

Inulin is a sugar found in dahlias and artichokes. It is small enough to be freely filtered by the kidney, but it is then neither reabsorbed nor secreted. The rate of *excretion* of inulin in the urine therefore depends

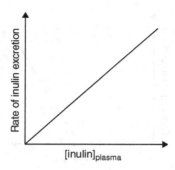

Fig. 80. The relationship between the rate of inulin excretion and plasma inulin concentration.

directly on the amount of inulin filtered per unit time. Assuming the rate of filtration is constant, it follows that the rate of excretion of inulin depends solely on $[inulin]_{plasma}$. We can see this by plotting the rate of inulin excretion against $[inulin]_{plasma}$ (Fig. 80) which gives a straight line (as plasma concentration increases, the rate of inulin excretion increases).

6.2.2 Glucose

Glucose, on the other hand, is freely filtered and actively reabsorbed by the tubules. As $[glucose]_{plasma}$ is increased, the rate of glucose filtration increases. Over the normal range of $[glucose]_{plasma}$, this increased rate of filtration is matched by a parallel increase in the rate of reabsorption. Therefore, in healthy subjects, no glucose is lost in the urine. However, as plasma glucose concentration is increased further, transport mechanisms become saturated and the rate of reabsorption can no longer increase to match the rate of filtration. Above a plasma glucose concentration of around 11 mM, glucose is then lost in the urine. The rate of this loss increases with increasing $[glucose]_{plasma}$ (Fig. 81).

6.2.3 PAH

Para-amino hippurate (PAH) provides an example of a substance which is freely filtered and then actively secreted by the tubules.

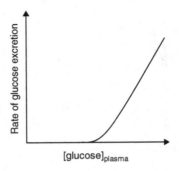

Fig. 81. The relationship between the rate of glucose excretion and plasma glucose concentration.

If the $[PAH]_{plasma}$ is kept low so as to avoid saturating the secretory mechanisms, the rate of excretion rises rapidly with increasing plasma concentrations. However, at higher $[PAH]_{plasma}$, the secretory mechanisms become saturated and the rate of excretion plateaus (Fig. 82).

We can further compare the handling of these substances by introducing the concept of *clearance*. The clearance of a substance is defined as the hypothetical volume of plasma from which the substance would have to be completely removed per unit time in order to produce the observed rate of excretion. A substance with a high clearance is therefore more effectively removed from plasma than one with a lower clearance. Anything which increases the renal excretion

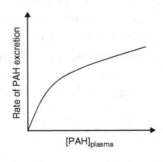

Fig. 82. The relationship between the rate of PAH excretion and plasma PAH concentration.

of a substance will tend to increase its clearance. If a substance is actively secreted by the tubules then its clearance will be greater than that of one that is only filtered. Conversely, if a substance is actively reabsorbed by the tubules then its clearance will be less than that of one that is only filtered. We can see this by comparing the clearance of our three substances.

Since inulin is neither reabsorbed nor secreted, the rate of excretion of inulin is directly related to the plasma concentration (Fig. 80). If the plasma concentration increases, so does the rate of excretion. The volume of plasma that would need to be completely cleared of inulin to account for the rate of excretion (the clearance) is therefore the same regardless of $[inulin]_{plasma}$ (Fig. 83).

Since glucose is actively reabsorbed, its clearance will be less than that of inulin. When $[glucose]_{plasma}$ is low, the transport mechanisms responsible for reabsorption are not saturated so no glucose appears in the urine and its clearance is zero. As $[glucose]_{plasma}$ increases, these mechanisms saturate and reabsorption makes proportionally smaller and smaller contributions. Glucose then behaves almost as if it was just freely filtered and the clearance approaches, but is always just less than, that of inulin (Fig. 84).

In contrast, since PAH is actively secreted, its clearance is greater than that of inulin. When $[PAH]_{plasma}$ is low, the transport mechanisms responsible for secretion are not saturated so the clearance is significantly greater than that of inulin. However, as $[PAH]_{plasma}$ rises

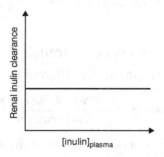

Fig. 83. The relationship between the renal clearance of inulin and plasma inulin concentration.

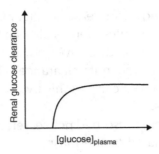

Fig. 84. The relationship between the renal clearance of glucose and plasma glucose concentration.

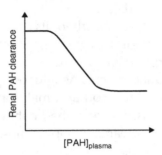

Fig. 85. The relationship between the renal clearance of PAH and plasma PAH concentration.

the transport mechanisms saturate and the proportionate contribution of secretion to the total excretion of PAH falls. The clearance of PAH then approaches, but is always just greater than, that of inulin (Fig. 85).

6.2.4 An alternative view of clearance

An alternative, less conceptual, definition of clearance is:

$$\text{clearance} = \frac{\text{rate of excretion}}{\text{plasma concentration}}$$

We can see how this definition arises from our earlier definition by again considering inulin, i.e. a substance that is freely filtered (the glomerular filtrate and the plasma contain inulin at the same concentration) and neither reabsorbed nor secreted. All the inulin that is excreted in the urine must therefore have arrived in the tubules as a result of ultrafiltration alone. If mass is to be conserved, then the amount of inulin filtered per unit time must equal the amount of inulin excreted per unit time and therefore:

volume of plasma filtered · plasma inulin concentration

= volume of urine excreted · urine inulin concentration

As all the inulin filtered is excreted and the concentration of inulin in the filtrate is the same as in plasma, the volume of plasma filtered is the same as the "hypothetical volume of plasma from which a substance would have to be completely removed per unit time in order to produce the observed rate of excretion".

We can therefore write:

$$\text{clearance} = \frac{\text{volume of urine excreted per unit time} \cdot [\text{inulin}]_{\text{urine}}}{[\text{inulin}]_{\text{plasma}}}$$

which simplifies to

$$\text{clearance} = \frac{\text{rate of excretion}}{\text{plasma concentration}}$$

Of course, most substances are not exclusively cleared from the plasma by filtration. However, this definition of clearance still applies. We can understand why with reference to inulin, glucose and PAH by considering situations in which the plasma concentrations of these substances are identical. It then follows that the same *amount* of each of the three substances will be removed from the blood at the glomerulus per unit time. Whilst all this inulin will be excreted, some (if not all) of the glucose is reabsorbed by the tubules and returned to the blood. Less glucose is therefore ultimately excreted and the rate of excretion of glucose is less than that of inulin. By the same argument, since PAH is secreted by the tubules and leaves the blood, more PAH is excreted and the rate of excretion is greater than that of inulin.

The amount of each of these substances cleared from the blood (excreted) therefore varies. As the concentrations of these three substances in plasma are the same, and:

$$volume = \frac{amount}{concentration}$$

it follows that the volume of plasma from which these amounts came from also varies. It is these "hypothetical" volumes of plasma that give the clearance.

Using this alternative definition, it can be seen that the clearance of a substance is given by the gradient of a plot of the rate of excretion against plasma concentration. We can see this by looking back at Figs. 80–85: the clearance at any plasma concentration (Figs. 83–85) is given by the gradient of the curve in Figs. 80–82 at that same concentration.

6.3   The Measurement of GFR

Before we leave the subject of clearance, we can also use it to understand how GFR is measured. Recall that GFR represents the volume of plasma which is filtered per unit time. The clearance of a substance that is freely filtered but neither secreted nor reabsorbed is also equal to the volume of plasma which is filtered per unit time. The clearance of a substance that is freely filtered but neither secreted nor reabsorbed therefore gives the GFR. We have already seen that inulin fulfils these requirements and it is in fact the usual physiologists' tool for experimental measurement of GFR.

$$GFR = \frac{\text{rate of inulin excretion}}{[inulin]_{plasma}}$$

GFR can be measured by infusing inulin at an increasing rate until $[inulin]_{plasma}$ becomes stable. At that point, the inulin infusion rate equals the inulin excretion rate. We therefore have all the information needed to calculate GFR without collecting urine at all. However, being able to measure GFR is not only of experimental interest but also of great clinical importance as decreases in GFR are often the first sign that renal failure is developing. From a clinical perspective, it would be far more convenient if GFR could be estimated without the need for an infusion of an exogenous substance. Cystatin C, a peptide produced by all nucleated cells at an apparently constant rate, is being evaluated for this purpose. At present, creatinine, the end product of the phosphocreatine intermediate energy system in skeletal muscle (see Chapter 2), is the most common measure used in clinical practice. Creatinine is secreted by the tubules, however, albeit at a low rate. Estimations based on creatinine clearance therefore all overestimate GFR. Incidentally, $[urea]_{plasma}$ may also be used to estimate GFR. However, this has the opposite problem: the tubules reabsorb urea at a low rate and so urea clearance underestimates GFR. With this proviso in mind, creatinine can be used to calculate GFR in the following equation:

$$GFR \simeq \frac{\text{rate of creatinine excretion}}{[\text{creatinine}]_{plasma}}$$

As creatinine is an endogenous compound, all that is needed for this is a plasma concentration of creatinine and a timed period of urine collection. An even more convenient, but less accurate, approximation of GFR can be obtained simply by measuring $[\text{creatinine}]_{plasma}$. Assuming that creatinine is produced at a constant rate, if the concentration in plasma increases, the rate of excretion and, by implication, the rate of filtration (GFR), must be decreasing (Fig. 86). In practice, the rate of creatinine production varies between individuals depending

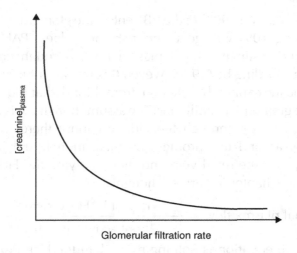

Glomerular filtration rate

Fig. 86. The relationship between plasma creatinine concentration and glomerular filtration rate.

on muscle mass and so formulae have been developed to take into account the variations with age, sex and race. It should be noted, however, that an increase in [creatinine]$_{plasma}$ beyond the normal range may not be seen until GFR falls to less than half its normal value. A raised [creatinine]$_{plasma}$ therefore usually indicates glomerular disease, while a normal plasma creatinine concentration does not exclude the presence of a glomerular problem.

6.4   The Measurement of Renal Plasma Flow

We saw above that PAH provides an example of a drug which is both freely filtered and actively secreted by the tubules. If [PAH]$_{plasma}$ is low enough to avoid saturating the secretory mechanisms, it can justifiably be assumed that PAH is completely cleared as it passes through the kidney. Just as the clearance of inulin can be used to estimate GFR, the clearance of PAH can then be used to estimate renal plasma flow.

$$\text{renal plasma flow} \simeq \frac{\text{rate of PAH excretion}}{[PAH]_{plasma}}$$

This can then be adjusted to give RBF as described in Chapter 5 (Further Thoughts).

In fact, only around 90% of the RBF enters the glomerular capillaries, the remaining 10% will go to nourish the kidney. PAH clearance therefore underestimates renal plasma flow. A rough correction is achieved by dividing by 0.9. However, this reveals a further problem: PAH obviously cannot be cleared from blood that does not pass through the glomerular capillaries. The assumption that it is completely cleared as it passes through the kidney cannot therefore be valid. We can get around this problem by measuring [PAH] in both the renal artery and the renal vein and then applying the Fick Principle (described in Chapter 5, Further Thoughts). Thus:

$$\text{renal plasma flow} = \frac{\text{rate of PAH excretion}}{[PAH]_{renal\ artery} - [PAH]_{renal\ vein}}$$

The clearance equation is nothing more than the Fick Principle with venous concentrations assumed to be zero.

Clinical Box 16: *Renal Replacement Therapy*

If GFR drops to a critically low level (around 10 ml·min^{-1}), continued survival depends on the removal of toxins by artificial means. This is achieved through *dialysis*. In *peritoneal dialysis*, a patient has a plastic tube permanently inserted into the abdomen which allows them to add and remove dialysate fluid to and from the intraabdominal cavity. Patients typically put warmed 2 L bags of varying concentrations (typically 80 mM–240 mM) of glucose into their peritoneal cavity. They then change them either approximately four times each day (*continuous ambulatory peritoneal dialysis*) or have a machine that changes them while they sleep at night (*automated peritoneal dialysis*). In haemodialysis, the patient is connected for several hours each week to a pump which removes blood, passes it through a dialysing machine and then returns it to the patient.

Despite the confusing nomenclature, both these clinical forms of "dialysis" involve *dialysis* and *ultrafiltration*. These can be varied independently according to the specific condition of the patient.

Dialysis is the diffusion of solutes across a semi-permeable membrane down a concentration gradient. It is the main method used to remove urea, creatinine and other waste products. In peritoneal dialysis, the *peritoneal membrane* that surrounds the intestine takes the role of the semipermeable membrane, while in haemodialysis, a synthetic material is used. In both cases, the concentration gradient arises because the dialysate fluid does not contain urea, creatinine or other waste products.

As we know, ultrafiltration describes the flow of water and other substances down a pressure gradient. It is primarily used to remove excess water. In peritoneal dialysis, the pressure gradient is the osmotic pressure gradient due to glucose, while in haemodialysis, it is a hydrostatic pressure produced by generating a negative hydrostatic pressure in the dialysate compartment.

6.5   Control of Plasma Osmolarity

We saw at the beginning of this chapter that it is extremely important to control the plasma osmolarity. Although there is a small contribution from thirst, this is achieved principally by varying the reabsorption, and hence excretion, of water. If the plasma osmolarity is too high, more water is reabsorbed to dilute the solutes in the plasma and decrease the osmolarity. Conversely, if the plasma osmolarity is too high, less water is reabsorbed and the solutes in plasma become more concentrated. As well as altering the osmolarity of the plasma, varying the water reabsorption also has the opposite effect on the osmolarity of the urine. Asking how the body controls plasma osmolarity is therefore the same as asking how the body controls the urine osmolarity. To answer this we need to look at how water is handled under normal conditions by the kidney and then how changes in osmolarity are detected and ultimately are able to alter water handling.

6.5.1   Water Handling by the Kidney

The first important concept to appreciate when considering the handling of water by the kidney is that all water reabsorption is passive. There are no water pumps. Water simply moves in response to osmotic gradients that result either directly or indirectly from the movement of solutes, mostly Na^+. If all regions of the kidney were equally permeable to Na^+ and water, the amount of water reabsorbed would simply reflect the amount of Na^+ reabsorbed and the osmolarity of the urine would reflect the osmolarity of the plasma. In order to be able to alter urine osmolarity independently, the kidney needs to be able to independently vary the reabsorption of Na^+ and water. This is achieved by having regions of the kidney which are able to reabsorb Na^+ without water and others which are able to reabsorb water without Na^+.

The first of these is the loop of Henle, which consists of three regions that function together to reabsorb more NaCl than water (Fig. 87). The overall result of this is to increase the osmotic pressure of the fluid in the medullary space and decrease the osmolarity of the tubular fluid. The fluid leaving the loop of Henle and entering the collecting duct is

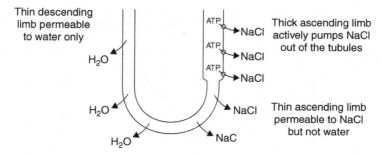

Fig. 87. The loop of Henle.

therefore always hypo-osmotic with respect to plasma. It is in the distal tubule and collecting duct that the regulation of water reabsorption occurs. Central to this is a substance known as *anti-diuretic hormone* (ADH) which is secreted from the posterior pituitary gland in the brain. In the absence of ADH, the collecting duct is impermeable to water. The osmolarity of the fluid entering the collecting tubule is therefore largely unchanged as it passes through the collecting tubule and hypo-osmotic urine is excreted. In the presence of ADH, however, pre-formed water channels (*aquaporins*) insert into the luminal membrane. The collecting duct therefore becomes permeable to water. With maximal ADH secretion, water can therefore be reabsorbed from the collecting tubule and the fluid can equilibrate with the medullary fluid. By varying the secretion of ADH from the posterior pituitary, the kidney can therefore vary the osmolarity of the urine.

This variation in secretion of ADH is controlled by osmoreceptors within the brain that detect changes in plasma osmolarity. These osmoreceptors are part of a projection (the *organum vaculosum of the lateral terminalis*) that pokes out across the blood brain barrier. Changes in plasma osmolarity result in the movement of water into or out of the cells and this increases or decreases the stretch on their membranes, opening or closing stretch-activated ion channels. In this way, they are sensitive to very small (1%–2%) changes in osmolarity and respond by sending signals to the pituitary gland to secrete ADH. If plasma osmolarity increases, more ADH is secreted whereas if plasma osmolarity decreases, less ADH is secreted (Fig. 88).

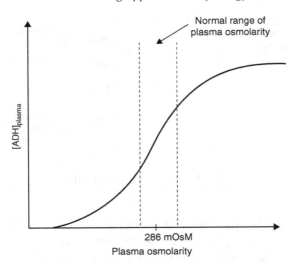

Fig. 88. The relationship between the plasma ADH concentration and plasma osmolarity.

So far so good. We have a sensor in the brain that detects changes in osmolarity and alters the rate of secretion of ADH. This alters the permeability of the collecting duct to water and so varies the osmolarity of the urine. However, it is not quite that simple. Recall that the movement of water is always passive. It follows that the osmolarity of the fluid in the tubules, and therefore of the urine, can never be greater than that of the medullary interstitial fluid with which it equilibrates. The production of urine with a high osmolarity in the presence of ADH therefore relies on the ability to generate a high osmotic pressure in the medulla. We saw earlier how the loop of Henle increases the medullary osmolarity by reabsorbing more NaCl than water. Were it as simple as this, increasing the permeability of the collecting duct to water would, at best, produce urine isosmotic with plasma. It would simply allow water that was not reabsorbed with NaCl in loop of Henle to be reclaimed. Furthermore, there is a maximum osmolarity gradient, around 200 mOsM across the tubular membranes because of leakage of water and back-diffusion of sodium. The generation of a hyper-osmotic urine therefore requires additional mechanisms — *counter-current multiplication* and *urea cycling*.

6.5.2 *Counter-current multiplication*

Since the tubule folds back on itself (hence the *loop* of Henle), changes in Na^+ and water movement in one segment can affect Na^+ and water movement in another. This allows the transverse gradient across the tubular wall to be multiplied into a much larger longitudinal gradient. Central to such *counter-current multiplication* is the fact that different regions of the loop of Henle have different permeabilities to Na^+ and to water. The thin descending limb is freely permeable to water but impermeable to Na^+. In contrast, the thin ascending limb is freely permeable to Na^+ but impermeable to water. Finally, the thick ascending limb contains pumps which actively transport Na^+ from the tubular fluid into the interstitium against the concentration gradient.

Although slightly counterintuitive, it is easiest to understand how these segments generate a longitudinal gradient by starting at the thick ascending limb. Imagine a situation where the tubular fluid is isosmotic with the interstitium (remembering though that it is unlikely that this situation would ever exists). The thick ascending limb actively pumps Na^+ from the tubular fluid into the interstitum. This raises the osmolarity of the interstitum and decreases the osmolarity of the tubular fluid until the maximum 200 mOsM difference in osmolarity between the tubular fluid and interstitum is reached. This creates an osmotic pressure gradient for water to diffuse out of the thin descending limb into the interstitum. The movement of water out of the tubular fluid without an accompanying movement of Na^+ increases $[Na^+]$ in the tubular fluid as it reaches the tip of the loop. Na^+ then passively moves from the fluid in thin ascending limb down its concentration gradient into the interstitium. In addition, more Na^+ can be actively pumped out of the thick ascending limb without exceeding the maximum 200 mOsM transverse gradient. Together, these increase the osmolarity of the interstitum, increasing the flow of water out of the thin descending limb. This generates a positive feedback cycle, further concentrating the fluid in the tip of the loop and increasing the movement of Na^+ out of the ascending limb. Throughout, the osmotic gradient driving water reabsorption is at its greatest when tubular fluid arrives at bottom of the loop of Henle. As the fluid moves into the ascending limb, Na^+ moves into the interstitium. This results in the magnitude of the osmotic gradient decreasing. A longitudinal gradient in [Na$^+$], and therefore in osmolarity, is thereby established.

A similar multiplication effect is achieved as a result of flow of fluid through the tubules. Movement of the hyperosmotic fluid in the tip of the loop of Henle up into the ascending limb itself allows the thick ascending limb to pump more Na$^+$ into the interstitum before the 200 mOsM maximum difference is reached.

6.5.3   Urea cycling

As a result of this counter-current multiplication, the 200 mOsM maximum transverse gradient is converted into a steep longitudinal gradient. However, the maximum osmolarity of urine that can be generated using this method alone is only 400 mOsM. Production of a more hyperosmotic urine requires the cycling of urea.

Urea is produced in the liver as the end product of nitrogen metabolism. It is freely filtered at the glomerulus and then progressively concentrated as water and other solutes are drawn out of the tubular fluid. By the time fluid reaches the collecting duct [urea] is considerably higher than in the interstitum. In the absence of ADH, all segments of the nephron other than the thin ascending limb are impermeable to urea. This urea is therefore excreted. However, in the presence of ADH the distal collecting duct is highly permeable to urea. Urea then diffuses down its concentration gradient out of the collecting duct into the interstitum. This itself raises the interstitial osmolarity and allows more water to passively leave the thin descending limb and more Na$^+$ to passively leave the thin ascending limb. Together, these effects further increase the osmolarity of the medullary interstitium and allow production of urine with an osmolarity of up to 1000 mOsM. In this way, when plasma osmolarity is high, urea is retained. This in effect allows Na$^+$ to be excreted in preference to urea, increasing the rate at which Na$^+$ is lost from the body. Once plasma osmolarity has been corrected, this urea must be excreted. This is where *cycling* comes in. When ADH is no longer present urea enters the tubular fluid in the thin descending part of the loop of Henle and, being unable to cross the distal collecting duct, is excreted in the urine. Were urea unable to re-enter the tubules, this toxic substance would remain trapped in the interstitium.

6.6 The Control of Extracellular Fluid Volume

The mechanisms described above control the solute concentration of the ECF largely by regulating the rate of excretion of water. Given that:

$$\text{volume} = \frac{\text{amount}}{\text{concentration}}$$

then, assuming that the solute concentration remains constant, ECF volume is directly related to the *amount* of solute in the ECF. Na^+ is the main ECF cation as it is so effectively excluded from cells and Cl^-, the main anion, follows Na^+ to maintain electroneutrality (see Chapter 1). Regulation of ECF volume is therefore tantamount to regulation of the amount of Na^+ in the body. While Na^+ intake is regulated to an extent, this is only usually important after severe volume depletion. As with most homeostatic processes, regulation is predominantly effected by balancing output against a variable input. It follows that ECF volume is predominantly controlled by controlling the rate of urinary excretion of Na^+. In this chapter, we will concentrate on the responses to changes in ECF volume that directly alter the rate of renal excretion of Na^+. It is important to remember, however, that changes in ECF volume include changes in circulating volume, and therefore produce changes in MSFP. Such changes alter VR and, by Starling's law, CO, ultimately altering ABP (see Chapter 5). This is important because, as we will see, it is through changes in ABP, rather than in $[Na^+]$ *per se*, that changes in ECF volume are detected. This also means that the responses to changes in ECF volume include the response to the resulting changes in ABP, as discussed in Chapter 7. In fact, long-term regulation of ABP is largely achieved through regulation of the total body Na^+ content.

Before discussing the detection of and responses to changes in ECF volume we must first consider the factors affecting the rate of Na^+ excretion. As we saw above, the rate of urinary excretion of any substance depends on the relative rates of filtration, reabsorption and secretion. We have already seen that Na^+ is freely filtered and then reabsorbed. Approximately two thirds of filtered Na^+ is reabsorbed in the proximal convoluted tubule, a further quarter in the loop of Henle and the remaining tenth-or-so in the distal collecting duct. The rate of urinary excretion of Na^+ can therefore be varied either by altering the rate at which it is filtered (the GFR) or the rate at which it is reabsorbed in any of these regions of the nephron.

6.6.1 Na^+ filtration

We saw above that GFR depends on both the blood flow through the glomerulus and the fraction of this blood that is turned into glomerular filtrate.

The blood flow through the glomerulus is determined by Darcy's law. We also know from the discussion in Chapter 4 that this longitudinal pressure difference is largely determined by ABP. Changes in ABP therefore have the potential to change blood flow through the glomerulus. As we have seen, autoregulatory mechanisms are in place to keep blood flow constant by balancing such changes in ABP with counteracting changes in arteriolar resistance. However, these mechanisms are not perfect and large changes in ABP can produce meaningful changes in the blood flow through the glomerulus.

The fraction of the blood flowing through the glomerulus that is turned into glomerular filtrate is also determined by Darcy's law. GFR therefore also depends on the transverse pressure gradient across the filtration barrier and the resistance to such flow. As we saw earlier, these parameters are described by the Starling filtration-reabsorption equation. Leaving aside small changes in the filtration coefficient that form a part of autoregulation (see above), GFR is controlled by controlling the hydrostatic pressure inside the glomerular capillaries. Since we wish to keep blood flow constant, this pressure can only be altered by altering the resistance to flow. As we saw above, resistance can be changed by changing vessel radius. The radii of the glomerular capillaries themselves cannot be changed since their walls do not include smooth muscle. However, the radii of the arterioles upstream (afferent arteriole) and downstream (efferent arteriole) can be altered. Constricting the afferent arteriole, and thereby reducing its radius, increases its resistance. This increases the pressure drop for each unit length along which blood travels before entering the glomerular capillaries. This decreases the pressure inside the glomerular capillaries and therefore decreases GFR. In contrast, constricting the efferent arteriole causes fluid to back-up in the glomerular capillaries, increasing pressure. This therefore has the opposite effect, increasing GFR. However, it is important to note that large increases in efferent arteriolar resistance inevitably decrease blood flow through the glomerular capillaries. In this situation, an increased fraction of a decreased volume of blood is filtered per unit time: the net result is a decrease in GFR. We will see below that several of the mechanisms for controlling Na^+ excretion exert their effects through constriction of these arterioles and the resultant changes in GFR.

6.6.2 Na^+ reabsorption

The mechanisms controlling the rate of Na^+ reabsorption are also rather involved. In the proximal convoluted tubule, Na^+ reabsorption is driven by *active transport*. This describes the movement of a substance against its concentration gradient and therefore necessitates the expenditure of energy. As with the majority of active transport processes in the body, Na^+ movement from the tubular fluid into the interstitium is driven by Na^+/K^+-ATPase pump (see Chapter 1). As we have established, this uses energy released from the hydrolysis of ATP to drive Na^+ movement. This generates a concentration gradient for Na^+ to move from the tubular fluid into the tubular cells. This is termed *secondary active transport* as an intermediate energy source is involved, with ATP being hydrolysed at a different site. Much of the Na^+ entering tubular cells is co-transported with glucose or amino acids, but an important part is carried on a Na^+/H^+ exchanger on which Na^+ enters the cells as H^+ leaves (Fig. 89). We will see later that two of the key signaling molecules that regulate Na^+ excretion act on this exchanger.

The rate of Na^+ reabsorption in the proximal tubule also depends on the rate at which Na^+ moves out of the peritubular fluid into the peritubular capillaries. If this rate of movement decreases, the

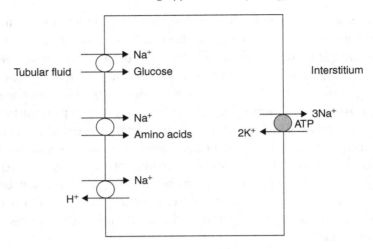

Fig. 89. Na$^+$ reabsorption in the proximal tubule.

pressure within this space (the *renal interstitium*) increases. This results in a decrease in the rate of Na$^+$ reabsorption from the tubules. Reabsorption in the proximal tubule is therefore directly affected by the factors appearing in the Starling filtration-reabsorption equation we encountered above.

In practice, the factors that vary to the greatest extent and have the greatest influence on the rate of Na$^+$ reabsorption are the hydrostatic pressure inside the peritubular capillaries and colloid osmotic pressure in the tubular capillaries. Through effects on these parameters, changes in ECF volume activate negative feedback loops within the proximal convoluted tubule that alter Na$^+$ excretion. Increases in the peritubular capillary hydrostatic pressure, such as occur as a result of increases in ECF volume and ABP, decrease movement of fluid into the peritubular capillaries and therefore decrease Na$^+$ reabsorption. The result is an increase in Na$^+$ excretion leading to a compensatory decrease in ECF volume and ABP. Similarly, decreases in the peritubular capillary colloid osmotic pressure, such as result from increases in ECF volume and dilution of plasma proteins, decrease the rate of movement of fluid into the peritubular capillaries. This decreases the rate of Na$^+$ reabsorption and hence decreases ECF volume.

This effect of variations of peritubular capillary colloid osmotic pressure leading to variations in Na^+ reabsorption is also made use of in the response to reductions in ECF volume. Many of the responses to reductions in ECF volume lead to preferential constriction of the efferent arteriole relative to the afferent. As described above, the result of this is not only a reduction in GFR but also an increase in the peritubular capillary colloid osmotic pressure as the filtration fraction increases and more fluid is drawn out of the blood. The result is an increase in Na^+ reabsorption in the proximal tubule.

Moving away from the proximal convoluted tubules, we saw above that the rate of Na^+ reabsorption in the loop of Henle is dependent on the medullary interstitial osmotic pressure, which in turn is determined by the plasma osmolarity. However, this is not the case for the Na^+ reabsorbed in the distal nephron. As in the proximal convoluted tubule, the reabsorption of Na^+ is driven by the Na^+/K^+-ATPase (Fig. 90). This transports Na^+ across the tubular epithelium into the peritubular fluid while Na^+ enters cells through epithelial Na^+ channels (ENaCs). We will see that changes in the activity and abundance of both ENaCs and Na^+/K^+-ATPases are both important means of altering the rate of Na^+ reabsorption.

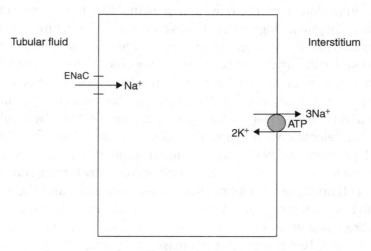

Fig. 90. Na^+ reabsorption in the distal tubule.

6.6.3 Varying Na⁺ excretion

Having discussed the various factors that affect the rate of excretion of Na^+, we are now in a position to consider how variations are brought about. Unlike the control of plasma osmolarity, the control of Na^+ excretion involves multiple overlapping pathways. The details of each of these pathways can appear overwhelming. However, they all affect Na^+ excretion by altering one or more of the factors described above.

The first of these involves *renin*, an enzyme secreted by modified smooth muscle cells in the wall of the afferent arteriole. Renin is secreted in response to decreased afferent arteriolar pressure and, via an increase in activity in the renal sympathetic nerves, in response to decreased ABP. Furthermore, a decrease rate of delivery of Cl^- to the macula densa also stimulates renin secretion. This makes the renin secretion sensitive to changes in GFR (if GFR falls less NaCl is filtered so less Cl^- reaches the macula densa), and NaCl reabsorption in the proximal tubule. The latter provides an unusual example of positive feedback. We will see that one of the end results of renin secretion is an increase Na^+ reabsorption in the proximal tubule. This decreases the rate of delivery of NaCl to the macula densa and stimulate further release of renin.

The importance of renin is that it stimulates the conversion of *angiotensinogen* to *angiotensin I*. This angiotensin I is in turn converted to *angiotensin II* by *angiotensin converting enzyme* found in the pulmonary capillaries. Angiotensin II then has a number of effects which together decrease the excretion of Na^+ and increase Na^+ intake: it stimulates Na^+ reabsorption in the proximal tubule via the Na^+/H^+ exchanger; it increases filtration of Na^+ by having a relatively selective vasoconstrictive effect on the efferent arteriole; it stimulates thirst and Na^+ appetite; and it stimulates the production of *aldosterone*. Aldosterone is a hormone secreted by the adrenal gland which acts on the distal nephron to increase the activity and abundance of both ENaCs and Na^+/K^+-ATPases. An increase in $[\text{aldosterone}]_{\text{plasma}}$ therefore leads to an increase in the rate of Na^+ reabsorption in the distal tubule. This results in a decrease in the rate of Na^+ excretion and a resultant increase in ECF volume.

Clinical Box 17: *Diuretics*

Diuretics are clinically important and frequently prescribed drugs used to increase the rate of urine production, decreasing ECF volume. This is particularly useful in heart failure, where decreasing ECF volume decreases preload and thereby improves cardiac function. It will be clear by now that diuretics must act by increasing the rate of Na^+, and therefore water, excretion. This is usually achieved by decreasing the rate of Na^+ reabsorption.

The most potent and clinically useful diuretics (*loop diuretics*) act to decrease the rate of Na^+ reabsorption in the loop of Henle. In this part of the tubule, Na^+ is reabsorbed on a $Na^+K^+2Cl^-$ carrier: drugs such as *frusemide* bind directly to the Cl^- site on this carrier, inhibiting it and thereby decreasing the rate of Na^+ reabsorption. The resulting increase in the osmolarity of the tubular fluid also decreases the rate of water reabsorption by osmosis, enhancing this effect. Importantly, these drugs not only increase the rate of excretion of Na^+ and Cl^-, but also of K^+. Patients receiving these drugs must be regularly monitored for decreases in $[K^+]$ plasma (*hypokalaemia*) as this can result in life-threating cardiac arrhythmias (see Chapter 1). Thiazide diuretics such as *metolazone* act by a similar mechanism, inhibiting the Na^+Cl^- carrier used for the reabsorption of solute in the distal tubule.

Other diuretics, including *spironolactone* and *amiloride*, act by blocking the effect of aldosterone. These do not have nearly such a potent effect as agents acting on other reabsorption mechanisms.

Osmotic agents such as *mannitol* constitute an interesting though potentially dangerous class of diuretics. These pharmcologically inert substances are freely filtered at the glomerulus but not reabsorbed by the tubules. They therefore draw water from the interstitium into the tubular fluid by osmosis, predominantly resulting in the loss of water. These drugs may be useful in acute renal failure when GFR is greatly reduced and tubular reabsorption of solutes becomes almost complete. These agents retain water in the tubular fluid and thereby ensure that urine flow does not completely cease.

It was mentioned above that a decrease in blood volume or ABP stimulates renin secretion via the renal sympathetic nerves. Increased activity of the renal sympathetic nerves also directly stimulates Na^+ reabsorption in the proximal tubules and constricts the renal arterioles, decreasing GFR and promoting Na^+ reabsorption.

An increase in ECF volume will lead to a decrease in activity in these renal sympathetic nerves and a decrease in renin secretion. In addition, *atrial natriuretic peptide* (ANP) is secreted in response to increases in ECF volume from the myocytes in the atria of the heart. This acts to inhibit Na^+ reabsorption in the proximal tubule and the collecting duct, inhibit the secretion of renin, aldosterone and ADH and dilate the renal arterioles.

6.7   The Responses to Intravenous Infusion of Distilled Water, Isotonic Saline and Hypertonic Saline

We have seen above how the body responds to changes in plasma osmolarity and ECF volume. Here we will look at how these control mechanisms interact by considering responses to the rapid intravenous infusion of 1 L of distilled water (a hypotonic fluid), isotonic saline and hypertonic saline. In order to compare the response to these three fluids, we first need to understand how they each distribute within the body. Under normal conditions, approximately two thirds of the body water is intracellular and the remaining one third is extracellular, either in the interstitial fluid, blood plasma, or transcellular fluid (see Fig. 91).

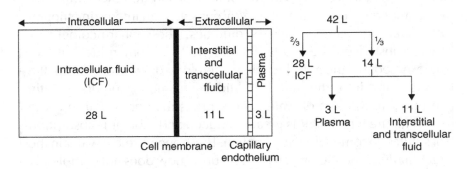

Fig. 91.   Distribution of body fluids.

Distilled water is able to distribute throughout the total body water. Therefore each compartment will receive fluid in proportion to its contribution to the total body water: approximately two thirds of the distilled water (0.67 L) will enter the ICF with one third (0.33 L) remaining in the ECF. In contrast, isotonic and hypertonic saline both contain Na^+ at a relatively high concentration. As we established in Chapter 1, resting cell membranes are largely impermeable to Na^+. This Na^+ will therefore be unable to enter cells and will osmotically retain its associated water. It follows that both isotonic and hypertonic salines remain entirely in the ECF. Therefore while infusion of any of these will increase the ECF volume, the degree of the change will differ. By a simple calculation, it can be shown that the infusion of distilled water will increase the ECF volume by only 2.4%, while the same volumes of isotonic and hypertonic saline will immediately each increase the ECF volume by 7.1%. Over time, however, hypertonic saline will draw water into the ECF from the ICF and so the increase will be even greater.

A change in volume of 2.4% is too small to be detected by the plasma volume-receptors: these require a 7%–10% before initiating a response. Hence, the infusion of distilled water will not trigger a volume-regulatory response, while those of isotonic and hypertonic salines will almost certainly do so. It is important to note at this stage that saline does not contain protein. Therefore as well as increasing ECF volume, these infusions will dilute the protein in plasma and decrease its colloid osmotic pressure. As established above, such a decrease in plasma colloid osmotic pressure will lead to an increase in GFR and ultimately an increase the rate at which Na^+ and water are excreted in the urine. This compounds the increased rate of Na^+ excretion brought about by other mechanisms.

We must now consider the effects of these infusions on plasma osmolarity. A simple calculation shows that the infusion of 1 L of distilled water (hypotonic) will decrease plasma osmolarity by approximately 2.5%, from around 286 mOsM to 279 mOsM. This change is certainly sufficient to be detected by the osmoreceptors, which are sensitive to a 1–2% change. Infusion of distilled water can therefore be treated as a purely osmotic problem, with a small added contribution from the change in colloid osmotic pressure resulting from the dilution of plasma proteins. By definition, an infusion of isotonic saline will not alter plasma osmolarity. In contrast, infusion of this same volume of hypertonic saline (let us assume an osmolarity of 1000 mOsM), will *increase* plasma osmolarity by 6.0% to 303 mOsM. Hence the infusion of hypertonic saline brings about both a volume-regulatory and an osmoregulatory response.

This brings up an important principle in homeostasis alluded to in the Introduction. Homeostasis is often described as the process by which the internal environment is kept constant. However, it is really more about minimising the overall effect of change on all the components that constitute the internal environment. Following infusion of hypertonic saline both ECF volume and plasma osmolarity increase: correcting one will inevitably compromise the other and therefore it would not be useful to attempt to correct both at the same time. The body therefore needs to decide which is more important. In this case, it is osmolarity, as even a small departure from the normal value can cause large changes in cell volume with potentially lethal effects on the brain as discussed earlier. Infusion of hypertonic saline therefore first results in the increased rate of secretion of ADH and of water reabsorption. In the absence of an accompanying increase in the rate of solute reabsorption, this results in dilution of the Na^+ in the plasma and returns osmolarity towards normal. Only once the plasma osmolarity is returned to near normal does the body begin to correct the increased ECF volume by excreting both Na^+ and water.

However, it cannot be the case that regulation of ECF osmolarity *always* takes precedence over regulation of ECF volume: this must depend on the severity of the challenge. This was vividly demonstrated in a heroic experiment by McCance in 1936. Over a period of several weeks, McCance and his colleagues ate diets almost entirely free from NaCl, as well as spending time in a sauna to cause sweating and hence further deplete NaCl from their bodies. As expected, their experiment (Fig. 92) showed that Na^+-loss was initially matched with water loss, so as to maintain plasma osmolarity constant: volume was sacrificed to maintain osmolarity. However, as the experiment continued and

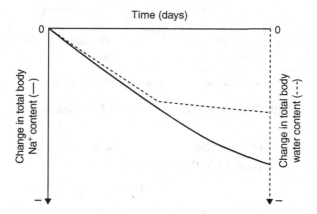

Fig. 92. Loss of Na^+ and water in McCance's experiment.

total body water volume fell by 8%–10%, this response was replaced by a drive to consume water: osmolarity was sacrificed to maintain volume.

FURTHER THOUGHTS

6.8   Osmotic Pressure, Osmolarity, and Tonicity

All substances will tend to move from a region of relative abundance (high concentration) to a region of relative scarcity (low concentration). This process is driven by a diffusive energy gradient, as described in Chapter 1. The movement of water from a region of high concentration to a region of low concentration across a membrane selectively permeable to water (*selectively-* or *semi-permeable membrane*) is given the term *osmosis*. This movement will result in a change in the relative volumes of water on each side of the membrane, causing it to stretch. Figure 93 illustrates such volume changes, showing the result when two glucose solutions of different concentrations are introduced into compartments separated by a membrane permeable to water but not to glucose.

This stretching occurs because a net force acts on the membrane. The force acting per unit area on each side of the membrane is the

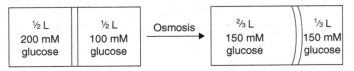

Membrane permeable to water but not glucose

Fig. 93. Movement of water by osmosis.

osmotic pressure. This can be calculated using van't Hoff's law, which takes similar form to the ideal gas equation:

$$\pi V = nRT$$

where π = osmotic pressure (Pa **or** mmHg)
V = volume (m^3 **or** L)
n = number of moles of particles which constitute *effective osmolytes* (mol **or** *Osmol*)
R = ideal gas constant (6.31 J·mol^{-1}·K^{-1})
T = absolute temperature (K).

The difference between the osmotic pressure acting on one side of the membrane and that acting on the other gives the net osmotic pressure. This net pressure determines the direction in which, and the degree to which, the membrane will stretch.

The van't Hoff equation can also be written as:

$$\pi = CRT$$

where C = concentration of all particles, or *effective osmolarity* (mol·m^{-3} **or** mol·L^{-1}).

The concentration of all particles (*osmolytes*) in the water (the *osmolarity*) therefore determines the osmotic pressure it is capable of exerting on the membrane. Osmolarity is a *colligative* property of the water and all the particles it contains: the identity of the particles does not matter. However, if the membrane is *permeable* to any of these osmolytes then they will not contribute to the osmotic pressure exerted: they are referred to as *ineffective osmolytes.* The concentration of *effective osmolytes* in the water (the *effective osmolarity*, usually measured in milliosmoles per litre, mOsmoles·L^{-1}) determines the osmotic pressure that it will exert. Effective osmolarity is also referred to as *tonicity*. Tonicity therefore depends both on osmolarity and membrane permeability. In other words, while osmolarity depends on the concentration of particles in water only, tonicity depends on both the concentration of particles in water and the permeability of the membrane.

Note that nowhere here do we use the term *solution,* as this implies that the osmolytes are dissolved in the water. They need not in fact be dissolved but can instead be *dispersed* as tiny particles spread evenly throughout the water to form a *colloid*. Plasma is an important example of such a colloid, consisting of water with dissolved ions and dispersed proteins. The osmotic pressure attributable to these protein components of plasma is referred to as the *colloid osmotic (oncotic)* pressure.

CHAPTER 7

Integrative Physiology

So far, we have considered each of the major systems in the body individually. However, much of physiology involves interactions between several of these systems. In this chapter, we consider the control of plasma pH, the control of arterial blood pressure and the response to exercise as three examples of such integrative physiology. In each case, we see how co-operative between these major systems facilitates homeostasis.

7.1   The Control of Plasma pH

Since H^+ is capable of becoming involved in a wide range of biological reactions, it is important that $[H^+]_{plasma}$ is held relatively constant. $[H^+]_{plasma}$ is kept between 36.3 nM and 45.7 nM by a number of interacting mechanisms involving the blood, kidneys, liver and lungs. These small concentrations are conventionally quantified as pH values, where:

$$pH = -\log_{10}[H^+],$$

where $[H^+]$ = concentration of H^+ ($mol \cdot m^{-3}$ **or** $mol \cdot L^{-1}$).

Hence, the normal plasma pH range is between 7.34 and 7.44. It should be noted that this logarithmic notation conceals the true range of permissible values: $[H^+]_{plasma}$ is allowed to vary by up to 20%. Nonetheless, the body is faced with the task of maintaining a relatively constant plasma pH despite varying inputs of H^+ from cellular metabolism and digestion. This control is not as straight-forward as simply varying the rate of urinary H^+ excretion directly. As we will see, much of the H^+ present in the body is bound to *buffers* and so is not available for excretion as free H^+. Furthermore, the minimum urine pH that can be created by active H^+ secretion is not especially low, being around 4.5 (about 30 μM). Given that the kidneys must facilitate the loss of around 3 mmol of H^+ each hour, a simple calculation shows that a urine production rate of 96 $L \cdot h^{-1}$ would be needed to accommodate the direct loss of this acid! Instead, control is achieved as a result of the coordinated properties and actions of the blood, kidneys, liver and lungs.

Central to all this is the concept of a buffer system. A buffer system is one that resists changes in pH following the addition of small amounts of acid or alkali. This property arises from the fact that in most buffer systems there is an equilibrium reaction between a weak acid (HA), which is able to donate H^+, and its conjugate base (A^-), which is able to accept H^+.

$$HA + H_2O \rightleftharpoons H_3O^+ + A^-$$

Note that H_3O^+ is effectively equivalent to H^+. If H^+ is added to the system, the conjugate base (A^-) accepts it to become HA and the equilibrium moves to the left. If the concentration of H^+ decreases, the opposite occurs: HA dissociates into A^- and H^+ and the equilibrium moves to the right. Small amounts of H^+ can therefore be added or removed from the system without producing a significant change in free $[H^+]$, or pH.

Just as for all other equilibrium reactions, we can write an equation for the equilibrium constant for a general buffer system:

$$K_a = \frac{[A^-] \cdot [H_3O^+]}{[HA] \cdot [H_2O]}$$

where K_a = equilibrium constant (dimensionless).

If we assume that there is an excess of water then we can ignore $[H_2O]$. Rearranging, this gives us:

$$K_a = [H_3O^+] \cdot \frac{[A^-]}{[HA]}$$

Since

$$pH = -\log[H^+] \quad \text{and} \quad pK_a = -\log K_a$$

taking logarithms of both sides to give

$$-pK_a = -pH + \log\frac{[A^-]}{[HA]}$$

or

$$pH = pK_a + \log\frac{[A^-]}{[HA]}$$

This is the *Henderson-Hasselbalch equation*. It can be seen from this that as long as the ratio of $[A^-]$ to $[HA]$ remains constant, pH will remain constant. We will see that this feature of buffer systems is central to the control of plasma pH.

There are three main buffer systems in the blood: carbonic acid and sodium bicarbonate; sodium dihydrogen phosphate and disodium hydrogen phosphate; and plasma proteins and their sodium salts. There are also similar buffer systems in red blood cells, but in this case K^+ rather than Na^+ is the cation (recall that Na^+ is largely excluded from cells) and the protein buffer is Hb and its salt. However, H^+ is unable to cross cell membranes directly or rapidly so these intracellular buffers are not involved in short term responses to changes in plasma pH.

Of the buffer systems in blood, the bicarbonate system is present in the largest quantity and is the main determinant of plasma pH.

$$H_2O + CO_2 \rightleftharpoons H_2CO_3 \overset{CA}{\rightleftharpoons} H^+ + HCO_3^-$$

When excess H^+ is present, the equilibrium is driven to the left and H^+ combines with HCO_3^- to form H_2CO_3. In the presence of the enzyme *carbonic anhydrase* (CA), this rapidly breaks down to form CO_2 and H_2O. This CO_2 is excreted by the lungs while the H_2O is excreted by the kidneys. Conversely, when $[H^+]$ decreases, H_2CO_3 breaks down into H^+ and HCO_3^-. In both cases these reactions normalise $[H^+]$, and therefore pH.

Using the same logic as above, we can write a Henderson-Hasselbalch equation for this buffer system:

$$pH = pK_a + \log\frac{[HCO_3^-]}{[CO_2]}$$

Substituting in the value for the pK_a of this system and replacing $[CO_2]$ with the product of solubility and partial pressure (see Chapter 4) we have:

$$pH = 6.1 + \log\frac{[HCO_3^-]}{0.03 \cdot P_{CO_2}}$$

where P_{CO_2} = partial pressure of CO_2 (Pa **or** mmHg).

From this, we can see that as long as the ratio of $[HCO_3^-]$ and P_{CO_2} remains constant, pH will remain constant. Importantly, these variables are independent and so an abnormality in the value of one can be compensated by altering the value of the other. Controlling the pH of plasma is therefore all about maintaining the value of this ratio.

7.1.1   Control of P_{CO_2}

We saw in Chapter 4 (Further Thoughts) that the arterial partial pressure of CO_2 is inversely related to the alveolar ventilation rate:

$$P_{ACO_2} = \frac{\dot{V}_{CO_2} \cdot P_{atmos}}{\dot{V}_A}$$

where P_{ACO_2} = partial pressure of CO_2 in alveolar gas (Pa **or** mmHg)
$\dot{V}_{CO_2}$ = rate of CO_2 production ($m^3 \cdot s^{-1}$ **or** $L \cdot min^{-1}$)
P_{atmos} = atmospheric pressure (Pa **or** mmHg)
$\dot{V}_A$ = alveolar ventilation rate ($m^3 \cdot s^{-1}$ **or** $L \cdot min^{-1}$).

We may assume, as we did in Chapter 4, that arterial plasma P_{CO_2} (P_{aCO_2}) reflects alveolar P_{CO_2}. It is therefore possible to control P_{aCO_2} by controlling the alveolar ventilation rate. A complex negative feedback system exists for this purpose. We will not deal with this in detail here but rather will focus on the relevant receptors. P_{aCO_2} is sensed by the *chemoreceptors*. These are found predominantly in two places: the *peripheral chemoreceptors*, which sample blood in the carotid arteries (*carotid bodies*) and the *central chemoreceptors*, which sample the cerebrospinal fluid (CSF) bathing the brain.

Ordinarily, the central chemoreceptors provide the most important input to the system controlling alveolar ventilation rate. This is despite their being on the brain side of the *blood-brain barrier*, a structure specialised to separate the brain's ionic environment from that of the body as a whole. Fortunately small, neutral molecules such as CO_2 are able to diffuse freely across the barrier and hence CSF P_{CO_2} reflects P_{aCO_2}. As CO_2 diffuses into these cells, H^+ is generated by the reaction described above. Interestingly, the protein content of the CSF is lower than that of the plasma and hence it is a less good buffer. This means that a particular change in P_{CO_2} will produce a larger change in pH in the CSF than in the plasma. This H^+ then inhibits cell membrane K^+ channels that ordinarily carry outward currents. Inward currents resulting from Na^+ leak into the cell then result in depolarisation (see Chapter 1), triggering signals in the afferent nerves linking with the brain.

If P_{aCO_2} rises relative to HCO_3^-, disturbing the ratio mentioned above, these central chemoreceptors trigger an increase in alveolar ventilation rate. This blows off additional CO_2 and so brings the arterial P_{CO_2}, and hence the HCO_3^- to P_{CO_2} ratio and arterial pH, back to normal. The opposite effect occurs if the ratio is disturbed such that P_{CO_2} decreases relative to HCO_3^- so that respiration is depressed. In this way, even relatively large changes in acid and base concentrations can be compensated for by changes in the respiratory rate. This feedback system is very sensitive: mild and scarcely perceptible adjustments of respiratory rhythm are probably taking place all the time in daily life.

A persistent increase in CSF P_{CO_2}, leading to a persistent decrease in pH, triggers an increase in the rate of HCO_3^- production by the *choroid plexus*, the collection of cells that produce CSF. Over the course of hours to days, the ratio of HCO_3^- to P_{CO_2} in the CSF can therefore be regulated independently from that in the plasma. This is particularly important on ascent to altitude where decreased atmospheric pressure results in decreases in both P_{CO_2} and P_{O_2}. The obvious solution to this decrease in inspired P_{O_2} would be to *increase* alveolar ventilation rate and so take in more O_2. However, arterial P_{CO_2} is already low and hyperventilation would cause a further decrease. The resultant increase in pH would, via these chemoreceptors, *decrease* the rate of alveolar ventilation. Initially, the ventilation rate oscillates in the presence of these two opposing effects. Over the course of hours to days, however, the rate of HCO_3^- production by the choroid plexus decreases to balance the decrease in P_{CO_2}. This corrects the ratio of HCO_3^- to P_{CO_2} in the region of the chemoreceptors and so the hypoxic drive from the peripheral chemoreceptors to increase ventilation rate is no longer blocked.

7.1.2 Control of $[HCO_3^-]$

In the short term, the buffer systems present in plasma, together with the rapid respiratory response to perturbations, are able to stabilise the pH. However, for each mole of H^+ removed as CO_2 and H_2O, a mole of HCO_3^- is lost. In order to return the buffer system to equilibrium, this HCO_3^- must be replaced. The kidneys have a key role to play in this process.

HCO_3^- is freely filtered in the glomerulus and under normal conditions all this filtered HCO_3^- is reabsorbed. However, HCO_3^- cannot be reabsorbed alone. Instead, it must first combine with H^+ ions in the tubular fluid to form H_2O and CO_2. This H_2O and CO_2 can then cross cell membranes into tubular cells, where they are converted back into H^+ and HCO_3^- (Fig. 94). In this way, the rate of HCO_3^- reabsorption is linked to the rate of H^+ secretion. It follows that the rate at which HCO_3^- is moved into the urine is determined by the difference between

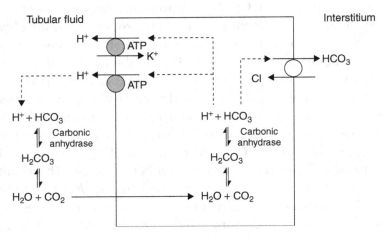

Fig. 94. The reabsorption of HCO_3^- in the distal nephron.

the rate at which HCO_3^- is filtered at the glomerulus and the rate at which H^+ is secreted by the tubules.

A closer look at this mechanism reveals that the H^+ that combine with HCO_3^- in the tubular fluid essentially cycle backwards and forwards across the membrane, carrying HCO_3^- into the cells. It is only when this H^+ combines with other buffers in the tubular fluid that it remains in the urine and is excreted. This link between HCO_3^- reabsorption and H^+ secretion explains how the kidneys deal with increased pH (*alkalosis*) and decreased pH (*acidosis*). In alkalosis, there is an excess of HCO_3^- in the filtrate. The kidneys respond by secreting too little H^+ to permit complete reabsorption of HCO_3^- and the excess is lost in the urine. In extreme alkalosis, *type-B intercalated cells* in the distal tubule and collecting ducts are also able to secrete HCO_3^-. However, there are usually relatively few of these cells and hence this mechanism is normally of little importance. In acidosis, the kidneys secrete enough H^+ to effect complete reabsorption of all filtered HCO_3^-. Any further H^+ secreted then combines with alternative buffers in the urine whilst additional HCO_3^- ions are generated to replace those lost as CO_2 via the lungs.

The means by which HCO_3^- is regenerated highlights a further important interaction, this time between the liver and the kidneys. The liver responds to acidosis by increasing the rate of glutamine production at the expense of the rate of urea production. It is believed that this switch occurs because the enzyme required for glutamine formation is pH-sensitive, though the underlying mechanism is still unclear. The kidneys then metabolise glutamine within the tubular cells to produce NH_4^+ and HCO_3^-. The NH_4^+ dissociates to from H^+ and NH_3. This H^+ is then pumped into the tubular fluid and NH_3 diffuses through the cells in the same direction. In the low pH in the tubular fluid, the two then recombine to form NH_4^+ which is excreted in the urine. Note that if the H^+ were not pumped into the tubular fluid, NH_4^+ would remain trapped in the tubular cells, preventing further breakdown of glutamine into NH_4^+ and HCO_3^- in turn preventing reabsorption of HCO_3^-. This is shown in Fig. 95.

An alternative view holds that the glutamine metabolism produces NH_3 rather than NH_4^+ but this does not really make a difference

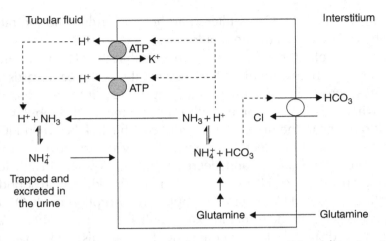

Fig. 95. The reabsorption of HCO_3^- and excretion of NH_4^+ in the distal nephron.

as it is still NH_4^+ that is lost in the urine. In both cases, it is the rate of H^+ secretion that ultimately controls the pH of the urine and the rate at which new HCO_3^- is added to the blood. It is not surprising then that H^+ secretion is triggered by a number of different stimuli. These include: an increased rate of filtration of HCO_3^-, a decreased plasma pH, an increased P_{aCO_2}, angiotensin II which stimulates Na^+/H^+ exchange and aldosterone which stimulates H^+ pumping into the tubules. Together, these stimuli enable close regulation of rate of H^+ secretion by the kidney, and ultimately of plasma pH.

7.2 The Control of Arterial Blood Pressure

We saw in Chapters 4 and 5 how the circulatory system serves a vital role in providing nutrients and removing wastes from each of the body's tissues. The demands of each of these tissues varies considerably with activity. It is therefore important to be able to independently control the flow of blood to each tissue. From Darcy's law, we know that:

$$\dot{V} = \frac{\Delta P}{R_{fluid}}$$

where $\dot{V}$ = flow ($m^3 \cdot s^{-1}$ **or** $L \cdot min^{-1}$)

ΔP = pressure difference (Pa **or** mmHg)

R_{fluid} = resistance to fluid flow ($kg \cdot s^{-1} \cdot m^{-4}$ **or** $mmHg \cdot L^{-1} \cdot min$).

It follows that the flow to a particular tissue could be altered either by varying pressure, or resistance, or both. We can imagine two extreme situations: one in which pressure is varied while resistance is kept constant and the other in which resistance is varied while the pressure is kept constant. To a first approximation, we can model the systemic circulation as a parallel circuit, with a single pump producing a pressure difference to drive flow. As we see in the Appendix, this means that the pressure difference across each organ is the same and also equal to the pressure difference around the entire circuit. It follows that, were the circulatory system to operate with varying pressures and a constant resistance, changing the flow through any one organ would require a change in the ABP. This would change the flow through every organ, making it impossible to change the relative distribution of flow between organs. Furthermore, increasing flow through the pulmonary circulation, for example during exercise, would mean increasing the pulmonary ABP. This would inevitably result in pulmonary oedema by the Starling filtration-reabsorption mechanism (see Chapter 5).

In contrast, the alternative option, controlling flow by varying resistance whilst keeping pressure constant, allows flow to individual organs arranged in parallel to be varied independently. Not surprisingly, this is the option employed by the circulatory system. By varying the resistance of the vessels supplying an organ, the flow through that organ can be varied without significantly altering flow to other organs. As long as the resistance to flow to all organs is not increased or decreased simultaneously, changes to the flow through individual organs can be achieved without substantially altering the total resistance of the circulatory system. To facilitate this, the control of the resistance to each organ varies depending on the relative importance of the organ. The resistance of vessels supplying the heart and brain depends on local activity. Hence, these vital organs are able to regulate their own blood flow in an activity dependent manner. Less immediately vital organs, such as those of the digestive system, have some capability to regulate their flow in much the same way. However, blood can also be diverted away if it is required to meet the urgent demands of more important organs. Even less important organs, such as the skin, have little regulation over their own supply and instead are completely at the mercy of whole body homeostasis.

Functioning as a constant pressure — variable flow system therefore enables the circulation to regulate flow to individual organs depending on their activity and relative importance. However, it does also mean that the heart and blood vessels are specifically adapted to work at constant pressure. Changes in ABP therefore have significant consequences for the entire circulatory system. As we saw in Chapter 5, increases in ABP increase the after-load against which the heart must pump and so put additional stress on the heart. Sudden increases in ABP can also cause small blood vessels to rupture, while long term increases result in damage the lining of blood vessels, predisposing to *atherosclerosis* amongst other problems. Conversely, decreases in ABP result in underperfusion of vital organs. Control over ABP is therefore vital to homeostasis. In fact, the circulatory system relies on the control of ABP to such an extent that its control is often at the expense of other, seemingly important, variables. We can see which these variables are by drawing together some of the equations we derived in Chapter 5:

$$\overline{\text{ABP}} = \text{CO} \cdot \text{TPR} \quad \text{where} \quad \text{CO} = \text{HR} \cdot \text{SV} = \text{VR} = \frac{\text{MSFP} - \text{RAP}}{\text{RVR}}$$

where $\overline{\text{ABP}}$ = mean arterial blood pressure (Pa **or** mmHg)
        CO = cardiac output ($m^3 \cdot s^{-1}$ **or** $L \cdot min^{-1}$)
        TPR = total peripheral resistance ($kg \cdot m^{-2} \cdot s^{-2}$ **or**
             $mmHg \cdot L^{-1} \cdot min$)
        HR = heart rate (beats min^{-1})
        SV = stroke volume (m^3 **or** L)
        VR = venous return ($m^3 \cdot s^{-1}$ **or** $L \cdot min^{-1}$)
        MSFP = mean systemic filling pressure (Pa **or** mmHg)
        RAP = right atrial pressure (Pa **or** mmHg)
        RVR = resistance to venous return ($kg \cdot m^{-2} \cdot s^{-2}$ **or**
             $mmHg \cdot L^{-1} \cdot min$).

It follows that ABP can be controlled by compensatory changes in either the properties of heart (SV and HR), the resistance within the circulation (TPR) or the blood volume (MSFP).

The SV and HR are principally controlled by the balance of activity between the sympathetic and parasympathetic limbs of the autonomic nervous system. TPR is also largely regulated by the autonomic nervous system. We saw in Chapter 5 that the largest proportion of the resistance lies in the arterioles and further that this resistance is related to the radius of the vessels:

$$R_{fluid} \propto \frac{1}{r^4}$$

where R_{fluid} = the resistance (kg·s^{-1}·m^{-4} **or** mmHg·L^{-1}·min)
  r = radius of pipe (m).

TPR can therefore be altered by widespread variations in the radius of the arterioles. This is achieved through sympathetic nervous system mediated constriction of smooth muscle in the walls of the arterioles: an increase in sympathetic nerve activity decreases the radius of the vessels and increases TPR.

We saw in Chapter 5 that MSFP is the pressure in the system at rest and is principally determined by the volume of blood in the arterial side of the circulation. MSFP can therefore be increased by either using the sympathetic nervous system to constrict veins in the splanchnic circulation and displace blood out of the venous side of the circulation, or by increasing the total circulating volume of blood.

7.2.1 Detection of changes in ABP

Before we see how changes in these variables are co-ordinated in response to changes in ABP, we need to consider briefly how changes in ABP are detected. The body has three complementary groups of receptors to do this.

Baroreceptors in the carotid sinus and the arch of the aorta act as stretch receptors and detect changes in ABP above 70 mmHg. By altering their firing rate they activate a negative feedback loop controlled by the brain and ultimately alter the activity of the autonomic nervous system. This loop is responsible for short term control, adjusting to daily variations and stabilising ABP. In a similar way, the chemoreceptors, which we discussed in Chapter 3 in reference to the control of P_{aO_2} and P_{aCO_2}, are stimulated by decreases in ABP below 60 mmHg and provide input to the same feedback loop. A third group of receptors, the volume receptors in the great veins, provide input to a second negative feedback loop concerned with the long term regulation of blood volume and therefore ABP (see later).

We can now see how these various factors interact by briefly considering the response to a decrease in ABP.

7.2.2   *Immediate control mechanisms*

In the short term, the effect of the decrease in ABP is minimised by variations in the all three of the above variables. Increased activity in the sympathetic nervous system, coupled with reciprocal decreases in activity in the parasympathetic nervous system, result in increases in both HR and SV. Increased sympathetic activity also results in arteriolar constriction, increasing TPR. Similarly, sympathetically-mediated venoconstriction in the splanchnic vascular beds displaces blood into the arterial side of the circulation. This increase is over the range of "stressed volumes" (Fig. 67) and therefore results in an increase in ABP.

7.2.3   *Intermediate term control mechanisms*

After a short delay, further changes in TPR and blood volume occur. An increase in TPR is achieved through an intrinsic property of blood vessels known as *reverse stress-relaxation*. Quite simply, when the pressure in the blood vessels becomes too high, the blood vessels stretch until pressure falls back towards normal (*stress-relaxation*). The converse, *reverse stress relaxation*, occurs as in this case with decreases in ABP. An additional increase in blood volume also occurs via a physical effect: the *capillary fluid shift mechanism*. A fall in the capillary hydrostatic pressure changes the Starling forces (Chapter 5) across the capillary wall and results in fluid being absorbed from the tissues, increasing blood volume. The opposite occurs with increases in ABP.

So far, all the mechanisms described are in place to minimise the effect of a decrease in ABP. If the decrease were due to a fall in blood volume, following haemorrhage for example, returning ABP to its normal level would require a compensatory *real* increase in blood volume. This is achieved via the kidney. A fall in ABP results in the release of renin. As we saw in Chapter 6, this increases the rate at which angiotensinogen is converted to angiotensin I, which is then converted to angiotensin II by *angiotensin converting enzyme* (ACE) in the lungs. Angiotensin II itself causes vasoconstriction and, together with aldosterone, renal retention of Na^+ and water. This explains why drugs that act on this system, including renin-inhibitors, ACE-inhibitors and angiotensin II-receptor blockers can be used to treat elevated ABP (*hypertension*).

7.2.4 *Long term control mechanisms*

Long term regulation of ABP in healthy individuals is all about regulation of blood volume. If blood volume is kept constant, ABP remains constant. The long term control mechanisms are therefore those that control ECF volume, and are therefore the ultimate responsibility of the kidney (Chapter 6). However, such regulation does not always result in a *normal* ABP. Patients in whom the blood vessels supplying the heart have become narrowed (*coronary artery stenosis*), for example, have no choice but to increase ABP if they are to maintain flow to this essential organ. In this case, the kidneys maintain an elevated ABP.

This explains why denervation of the baroreceptors leads to greater variability in ABP, but no change in the mean ABP. These baroreceptors provide the most important input to the negative feedback system responsible for the short-term control of ABP. Interestingly, denervation of these baroreceptors also eliminates low frequency oscillations in ABP (*Mayer waves*) that are seen in normal individuals. These may reflect oscillations in this negative feedback system, as discussed in the Appendix.

In contrast, denervation of the volume receptors does not affect ABP-variability but increases mean ABP. This is because the volume receptors provide the key input to the renal negative-feedback system responsible for the long term regulation of ABP, leading to fluid retention, increased blood volume and hence increased ABP.

7.3   The Response to Exercise

The ability to catch prey or escape from a predator is of clear evolutionary importance. Skeletal muscle is essential for this purpose, however, we also need a means of providing this muscle with nutrients, and removing waste products at a sufficient rate. For example, while the resting rate of O_2 consumption in a human subject is around $300\,ml\cdot min^{-1}$, it may increase by up to ten times to $3\,L\cdot min^{-1}$ during extreme exercise. This increased requirement must be achieved by the respiratory and circulatory systems.

Clearly, to achieve this, blood flow to skeletal muscle must increase during exercise. This cannot usually be achieved by diverting blood from other tissues. Instead an increase in CO is required. This may increase by up to six times during extreme exercise, facilitating the largest part of the increase in O_2 delivery to the tissues. Notably, in healthy individuals, it is limitations in the capacity to increase CO that limits maximum exercise performance. We saw in Chapter 6 that:

$$CO = HR \cdot SV$$

where HR = heart rate (s^{-1} **or** min^{-1})
   SV = stroke volume (m^3 **or** L).

Increasing HR, SV, or both will therefore increase CO. However, as we have seen, this effect is limited. Without making other changes in the circulation, even replacing the heart with a high capacity pump can only increase CO by a maximum of 5%–10%: the heart can only pump out the blood that it receives. These larger increases in CO therefore require an increase in the rate at which blood is returned to the heart, the VR. The Starling mechanism then ensures that the heart is able to pump this additional blood, facilitating the increase in CO (Chapter 5). As well as increasing HR and SV, much of the response to exercise therefore centres around increasing VR.

Before we consider the specific changes that occur, it is helpful to remind ourselves of the factors that control VR:

$$VR = \frac{MSFP - RAP}{RVR}$$

where VR = venous return ($m^3 \cdot s^{-1}$ **or** $L \cdot min^{-1}$)

   MSFP = mean systemic filling pressure (Pa **or** mmHg)
   RAP = right atrial pressure (Pa **or** mmHg)
   RVR = resistance to venous return ($kg \cdot m^{-2} \cdot s^{-2}$ **or**
      $mmHg \cdot L^{-1} \cdot min$).

From this equation, it appears that decreasing RAP would increase VR. However, in reality, RAP is maintained at around 0 mmHg as any further decrease would generate a negative pressure in the veins entering the right atrium and cause them to collapse (Chapter 6). In fact, RAP *increases* slightly during exercise as a result of contraction of smooth muscle in the veins. Increases in VR can therefore be achieved through increases in MSFP and decreases in RVR. We will see that both such changes play a part in the response to exercise and in fact begin to occur before exercise begins.

7.3.1 Anticipatory responses

The anticipatory response to exercise provides an example of feed-forward within the cardiovascular system. This is necessary both because smooth muscle is much slower to contract than skeletal muscle (Chapter 2) and because changes in cardiac contractility take time to come into effect. The anticipatory response means that we do not have to wait for an error signal to be detected by the baroreceptors before the necessary adjustments are made.

Higher brain centres, acting via the hypothalamus and medulla, control these anticipatory responses which are largely mediated by the release of endocrine vasoconstrictors, including adrenaline and angiotensin II. The resultant contraction of smooth muscle in the walls of veins throughout the body squeezes blood into the arterial side of the circulation and effectively increases the circulating volume. As discussed further in Chapter 5, this increase in volume results in a small increase in MSFP. However, there is a limit to the degree to which such an increase in MSFP is useful. Applying the Starling filtration-reabsorption equation (Chapters 4 and 5), it can be seen that the rise in capillary hydrostatic pressure which accompanies a rise in MSFP will increase the rate at which fluid moves transversely across vessel walls into the tissues. The resultant increase in the volume of fluid in the tissues not only results in a decrease in circulating volume but also increases the distance between tissue and blood over which substances must diffuse by the Fick equation (Chapter 4). Even the relatively minor two to three times increase in MSFP that occurs causes a 10% drop in circulating volume.

Adrenaline and other hormones also increase the ability of the heart to pump blood as part of the anticipatory response: heart rate increases (positive *chronotropic effect*) and cardiac contractility increases (positive *inotropic effect*). There is some evidence from human studies that training can alter the magnitude of this response. In general, the shorter the anticipated duration of exercise, the greater the anticipatory rise in heart rate. This makes sense since increasing heart rate is effective but metabolically expensive.

These increases in HR and SV increase CO. However, so far we have not considered the effects an increased CO has on the rest of the circulation. Recall that from Darcy's law:

$$CO = \frac{ABP}{TPR}$$

where ABP = arterial blood pressure (Pa **or** mmHg)
TPR = total peripheral resistance (kg·m^{-2}·s^{-2} **or** mmHg·L^{-1}·min).

An increase in CO without an accompanying decrease in TPR must therefore increase ABP. We will see later than small changes in ABP do indeed occur, but these are limited by decreases in TPR. The reduction in TPR is brought about in this anticipatory phase by the opening of vessels known as *arteriovenous anastomoses*. These in effect short-circuit much of the resistance in the arterial component of the circulation, taking blood from the arteries and returning it directly to the veins. It is significant that TPR is lowered during the anticipatory response by opening these arteriovenous anastomoses and not by simply opening the capillary beds. On a purely practical note, the opening of capillary beds is controlled by the quantity of vasodilating metabolites present. It would not therefore be possible to open the capillary beds during the anticipatory response even if the body knew which beds to open. Increasing flow through the muscle vascular beds would also wash out any metabolites that had accumulated, producing compensatory vasoconstriction and hence an increase in resistance.

7.3.2   *Responses during exercise*

Increased muscle activity during exercise, leads to the local accumulation of a range of metabolites in the ECF. As alluded to above, many of these factors have a vasodilator effect, decreasing local arteriolar resistance and diverting blood flow through the muscle vascular beds. Indeed, during exercise, flow through muscle vascular beds is almost linearly related to local metabolic rate. Arteriovenous anastomoses can then be closed without increasing TPR. The build up of metabolites in the tissues further increases the blood flow through the muscle vascular beds by opening pre-capillary sphincters. This increase in the number of perfused capillaries also reduces the distance between capillaries and muscle cells, facilitating diffusion.

However, these effects do not in themselves account fully for the increases in VR during exercise. In fact, the largest contribution is made by the reduction in RVR brought about by *muscle* and *respiratory pumping* (Chapter 4). Briefly, as skeletal muscle contracts and relaxes it compresses the veins running through it, squeezing blood back towards the heart (backflow is prevented by valves). A similar pumping effect arises as a result of increased movements of the diaphragm as the rate and depth of breathing increases.

These effects are summarised in Fig. 96, which shows the changing venous return and cardiac output curves before, in anticipation of and during exercise.

In this way, the circulatory system is able to increase the transport of blood to and from the tissues during exercise. However, the increased demand for provision of O_2 and removal of CO_2 also requires an increase in the rate of gas exchange in the lungs. This is achieved

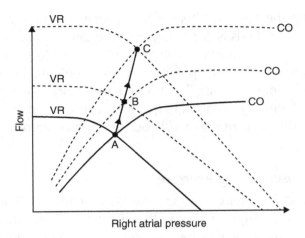

Fig. 96. Changes in cardiac and vascular function curves before and during exercise.

A → B : Anticipatory phase. Both contractility and MSFP increase.

B → C : During exercise. Contractility and MSFP increase further. RVR also decreases.

both by increasing the alveolar ventilation rate and by increasing the surface area available for gaseous exchange (see again the Fick equation in Chapter 4). Alveolar ventilation rate is primarily controlled by P_{aCO_2}, as we saw both earlier in this Chapter and in Chapter 4. The increased rate of CO_2 production during exercise will, therefore, itself increase alveolar ventilation rate. However, an increase in ventilation is seen immediately on beginning exercise, before P_{ACO_2} has had the opportunity to change and generate a feedback signal. This has led to suggestions that the early increase is brought about by direct communication between brain centres. It may be the case that the respiratory centres are directly activated as nerves to skeletal muscle are activated. Movement sensors in joints (*proprioceptors*) are also believed to play a part in this response as even passive movements often increase the rate and depth of breathing.

The increase in the surface area for exchange of gases occurs as a direct consequence of the increase in CO that occurs during exercise. As CO increases, flow through both the systemic and pulmonary circulations increases. Pulmonary perfusion therefore increases, recruiting more alveoli. This increase in pulmonary perfusion inevitably means that blood flows faster through the alveolar capillaries, leaving less time available for gas exchange. However, this effect is outweighed by the increase in surface area and gas exchange therefore increases.

7.3.3 Changes in ABP during exercise

Remarkably, despite these significant increases in CO, during normal forms of exercise ABP increases to a lesser degree than might be expected. This is because the increases in CO are in part balanced by decreases in TPR which occur as a result of the opening of muscle vascular beds. However, unusual forms of exercise that do not result in a significant muscle pump effect, for example holding a heavy weight steady for a long period of time, result in a different response. In this case, metabolites which accumulate in the muscle trigger sensory nerves and result in a nervous response that dramatically increases ABP. Conversely, patients with severe cardiovascular disease often experience a decrease in ABP during normal forms of exercise. In this case, the increase in CO that takes place is insufficient to compensate for the decrease in TPR.

Derivations and Theoretical Points

In this Appendix we explore a number of the key principles underlying physiology. We begin by discussing the properties of feedback systems, the very essence of homeostasis, in more detail. We then reflect on why it is that matter and charge flow from one place to another. Next, we develop this discussion to include the concept of *capacitance*. We then go on to consider two fundamental points which arise from this discussion, the measurement of pressure and LaPlace's law. Finally, we address two specific points arising from Chapter 1, namely the maintenance of cell volume and the propagation of signals along nerve axons.

Control in Physiology

Homeostasis, the maintenance of a constant internal environment independent of the external environment, is the very essence of physiology. Achieving this goal requires the body's systems to respond to externally imposed changes in such a way so as to return variables to their *set points*. Many physiological processes, such as the regulation of plasma osmolarity to name but one example, can be understood using a framework originating from engineering, *control theory*. The *controlled variable* (plasma osmolarity in this case) is sensed by *receptors* (mainly the hypothalamic osmoreceptors). These set up an *input signal* which is fed via the afferent nerves to a *controller* (the hypothalamus). This generates an *output signal* (increased production of ADH).

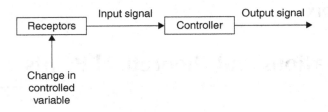

Fig. 97. Open-loop control system.

This system could be described as an *open loop* (Fig. 97): input affects output but output does not affect input. For this simple system, we can define the *amplification*, or *open loop gain*, A_o as:

$$A_o = \frac{\text{output}}{\text{input}}$$

On its own, this is not particularly useful. If plasma osmolarity is to be regulated then the output signal needs to produce a change in the controlled variable, i.e. the loop must be closed by a *feedback pathway*. Returning to our example, the output signal (ADH) acts on an *effector* (the kidney) to produce a change in plasma osmolarity. The system can then feedback on itself, in effect adding a component of the previous output to the new input. Notably, it is important that the time taken for the system to respond to a change in the controlled variable (the *latency*) be short in relation to the time course of the change that is to be corrected. If it is not then the system may respond inappropriately, potentially resulting in oscillations in the controlled variable.

The closed loop system shown in Fig. 98 can be described with the addition of a new parameter, the *feedback factor*, β. This is defined as the fraction of the output added to the input. The input to the controller in the next cycle (*input'*) is then:

$$\text{input}' = \text{input} + \beta \cdot \text{output}.$$

The effective amplification, or *closed loop gain*, A_c is then:

$$A_c = \frac{\text{output}}{\text{input}'}$$

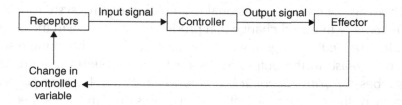

Fig. 98. Closed-loop control system.

or

$$A_c = \frac{\text{output}}{\text{input} + \beta \cdot \text{output}}$$

Rearranging, this gives:

$$\text{input} + \beta \cdot \text{output} = \frac{\text{output}}{A_c}$$

and dividing by *input*:

$$1 + \beta \cdot \frac{\text{output}}{\text{input}} = \frac{\text{output}}{A_c \cdot \text{input}}$$

or

$$1 + \beta \cdot A_o = \frac{A_o}{A_c}$$

which rearranges to:

$$A_c = \frac{A_o}{1 + \beta \cdot A_o}$$

If the system is to return to its set point after an externally imposed change then an increase in the output must result in a feedback signal which, when mixed with the input, results in a decrease in the output in the next cycle around the loop. This is tantamount to saying that A_c must be negative. Of course, A_o must always be positive (a negative amplification would be meaningless), so β must be less than -1 (naturally $A_o \neq -1$). If $\beta < -1$ then the system is a *negative feedback system*. It is important to remember that while A_o is a characteristic of the controller, β is a characteristic of the feedback pathway: it is not the hypothalamus that makes our example a negative feedback system but rather the response of the kidney to ADH.

There are, of course, occasional examples of control systems where an externally imposed change that produces an increase in the output results in a feedback signal which, when mixed with the input, results in an *increase* in the output in the next cycle around the loop. This describes a *positive feedback system*, defined by A_c being positive. Curiously, there are no theoretical constraints on β in such systems (β can in fact be negative if $A_c < 1$): either the controller or the feedback pathway may be responsible. If A_c is positive and < 1 then each successive cycle adds a smaller additional feedback signal and the value of the controlled variable settles to a new, larger, value. Such systems are occasionally found in physiology and the probability of opening of the voltage-activated Na^+ channel represents an important example. However, if $A_c > 1$ then each successive cycle adds a larger additional feedback signal and the value of the controlled variable explodes! After a large haemorrhage, the A_c of the system regulating arterial blood pressure may change from being negative (negative feedback) to > 1 (explosive positive feedback). If left unchecked, such a transition is rapidly fatal.

Flow, Resistance and Energy

It is a fundamental property of the universe that all systems tend to move towards their lowest possible energy state (this is the Second Law of Thermodynamics). From this simple principle, we can derive a number of fundamental equations describing flow, both of matter and of charge (electrical current). Here we will only consider situations where such flows are constant.

It should be noted from the outset that mass and charge represent comparable fundamental quantities. However, while mass has its own SI base unit (the kilogram), charge does not. Instead, charge is expressed as a derived unit (the coulomb), where $1\,C = 1\,A \cdot s$. This often leads to needless confusion.

The flow of fluids

Since the flow of matter is easier to visualise than the flow of charge, we will begin here. We could equally well describe such flow as the flow of mass, or as the flow of volume. It is usually more convenient in physiology to consider the flow of volume, rather than that of mass. The rate at which a fluid flows is given by:

$$\dot{V} = \frac{dV}{dt}$$

where $\dot{V}$ = flow ($m^3 \cdot s^{-1}$ **or** $L \cdot min^{-1}$)
V = volume (m^3 **or** L)
t = time (s).

If a constant force is acting then:

$$E = Fx$$

where E = work done or energy converted (J)
F = force (N)
x = distance (m).

Newton's Second Law states that for a constant acceleration:

$$F = ma$$

where m = mass (kg)
a = acceleration ($m \cdot s^{-2}$).

In a gravitational field, all masses are constantly being accelerated. The weight of an object is due to the action of this field on the object such that:

$$W = mg_g$$

where W = weight (N)
g_g = acceleration due to gravity ($m \cdot s^{-2}$).

As well as being the acceleration due to gravity, g_g is also the *force per unit mass* ($1 \ m \cdot s^{-2} = 1 \ N \cdot kg^{-1}$). In a gravitational field, *force per unit mass* is the definition of *field strength*.

So we have:

$$E = mg_gx \text{ (which might be more familiar as } PE = mgh).$$

Mass is the product of volume and density:

$$m = \rho V$$

where ρ = density (kg·m^{-3} **or** kg·L^{-1}), so combining the above two expressions we have

$$E = \rho Vg_gx$$

Stepping aside for a moment, let us think about the relationship between energy and pressure. If pressure is constant then we can think of force as the product of pressure and area:

$$F = PA$$

where P = pressure (Pa **or** mmHg)
A = area (m^2).

Going back to:

$$E = Fx$$

we can now write

$$E = PAx$$

and since the product of area and distance is volume:

$$E = PV$$

so

$$P = \frac{E}{V}$$

i.e. pressure represents energy per unit volume. This is of great importance in understanding the flow of blood through the circulatory system and the flow of gases through the respiratory system.

The flow of charge

Let us now think about the flow of charge. This is of relevance in understanding the electrical properties of nerve and muscle cells. Current is the flow of charge:

$$I = \frac{dQ}{dt}$$

where I = current $(C \cdot s^{-1}, A)$
Q = charge $(C, A \cdot s)$.

By analogy with Newton's second law, when a charge is placed in an electric field:

$$F = Qg_e$$

where Q = charge (C)
g_e = force per unit charge $(N \cdot C^{-1})$.

Incorporating this into our definition of energy as work done:

$$E = Qg_e x$$

In this case, $g_e x$ is the energy needed to move a charge Q a distance x in an electric field. This time we can write:

$$E = QV_e$$

where V_e = electrical potential energy difference per unit charge $(J \cdot C^{-1}, V)$

$$V_e = \frac{E}{Q}$$

i.e. *potential difference* is energy per unit charge. Pressure difference and potential difference therefore represent comparable quantities.

Flow in general

Just as with the flow of fluid, the flow of charge occurs from regions of high energy to regions of low energy. In this context, this means regions of high potential to regions of low potential, i.e. current flows because of a potential difference.

It is conventional to talk about field strength, g_g, in a gravitational field and potential energy difference per unit charge, V_e, in an electric field. Of course in both cases:

$$V = gx$$

This difference in convention needlessly causes confusion.

It follows that fluids flow from regions of high pressure to low pressure and that charge flows from regions of high potential to low potential. This is equivalent to saying that fluid or charge will flow from point a to point b if the pressure or potential at a is higher than that at b, i.e. flow, $\dot{Q}$, occurs because of a pressure or potential difference, Δy. In fact:

$$\dot{Q} \propto \Delta y$$

Introducing a constant of proportionality, g, we have,

$$\dot{Q} = g\Delta y$$

Here g represents *conductance*. However, it is more conventional to think of resistance, R where:

$$R = \frac{1}{g}$$

So we have:

$$\dot{Q} = \frac{\Delta y}{R}$$

For fluids, this is Darcy's law:

$$\dot{V} = \frac{\Delta P}{R_{fluid}}$$

where $\dot{V}$ = flow along the tube ($m^3 \cdot s^{-1}$ **or** $L \cdot min^{-1}$)
  ΔP = pressure difference driving flow (Pa **or** mmHg)
  R_{fluid} = resistance to flow of fluid ($kg \cdot m^{-2} \cdot s^{-2}$ **or** $mmHg \cdot L^{-1} \cdot min$)

and for charge is it Ohm's law:

$$I = \frac{\Delta V_e}{R_{charge}}$$

where R_{charge} = resistance to flow of charge (Ω).

It must be kept in mind that these laws only apply if certain conditions are met. For fluid flow, the fluid must be Newtonian, like water (this is defined by shear stress being directly proportional to the velocity gradient in the plane perpendicular to that in which the sheer occurs, but we will leave this point here). This effectively means that the fluid continues to behave in the same way no matter how quickly it flows (it has a constant *viscosity*). Flow must be *well-developed* (it is not valid just as fluid enters a pipe) and flow must be laminar (flow occurring in parallel layers with no interactions between the layers). We consider the extent to which blood in the circulatory system exhibits these properties in Chapter 4 (Further Thoughts). Finally, the pipe through which flow occurs must be rigid. For current, the resistor must be Ohmic, i.e. $I \propto \Delta V_e$.

We can extend the analogy between fluid flow and current further. If the above conditions are met and flow occurs along a cylindrical resistor, resistance to fluid flow can be calculated as:

$$R_{fluid} = \frac{8\eta l}{\pi r^4}$$

where η = viscosity (kg·m^2·s^{-1})
   l = length of tube (m)
   r = radius of tube (m)

R_{charge} is subsequently referred to as R while resistance to current can be calculated as:

$$R_{charge} = \frac{\rho l}{\pi r^2}$$

where ρ = resistivity (Ω·m)

Flow of matter or charge through a resistor of finite length results in a finite drop in pressure or potential, which can be calculated with these equations. This of course means that energy has been consumed in doing work against the resistance. This work done is given by:

$$E = V\Delta P \text{ for fluid flow}$$

$$E = Q\Delta V \text{ for charge flow (current)}$$

In both cases, an increase in the length of the resistor produces a proportionate increase in resistance. An increase in the radius of the

A Thinking Approach to Physiology

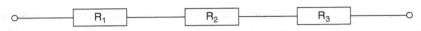

Fig. 99. Resistors in series.

resistor produces a large decrease in resistance with fluid flow being more affected than current. In both cases, resistance also depends on an additional constant of proportionality, viscosity in the case of fluid flow and resistivity in the case of current.

We can now begin to think about networks of resistors. In the *series* network (Fig. 99), mass and charge must flow through each resistor in turn, i.e. the flow through each resistor is the same.

So the potential difference or pressure difference across each resistor is given by:

$$\Delta V_1 = IR_1, \quad \Delta V_2 = IR_2, \quad \Delta V_3 = IR_3,$$
$$\Delta P_1 = \dot{V}R_1, \quad \Delta P_2 = \dot{V}R_2, \quad \Delta P_3 = \dot{V}R_3.$$

Assuming that energy is conserved in the circuit, i.e. the First Law of Thermodynamics applies, then:

$$\Delta V = \Delta V_1 + \Delta V_2 + \Delta V_3, \quad \Delta P = \Delta P_1 + \Delta P_2 + \Delta P_3$$

so

$$IR_{tot} = IR_1 + IR_2 + IR_3, \quad \dot{V}R_{tot} = \dot{V}R_1 + \dot{V}R_2 + \dot{V}R_3$$

and

$$R_{tot} = R_1 + R_2 + R_3, \quad R_{tot} = R_1 + R_2 + R_3$$

So the total resistance of a series network of resistors is simply the sum of each individual resistance.

In the *parallel* network shown in Fig. 100, there are three possible paths which flow can take. However, the potential difference or pressure difference across each path is the same.

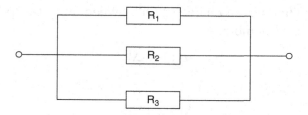

Fig. 100. Resistors in parallel.

Therefore:

$$I_1 = \frac{\Delta V}{R_1}, \quad I_2 = \frac{\Delta V}{R_2}, \quad I_3 = \frac{\Delta V}{R_3},$$

$$\dot{V}_1 = \frac{\Delta P}{R_1}, \quad \dot{V}_2 = \frac{\Delta P}{R_2}, \quad \dot{V}_3 = \frac{\Delta P}{R_3}$$

and

$$I_{tot} = I_1 + I_2 + I_3, \quad \dot{V}_{tot} = \dot{V}_1 + \dot{V}_2 + \dot{V}_3$$

so

$$\frac{\Delta V}{R_{tot}} = \frac{\Delta V}{R_1} + \frac{\Delta V}{R_2} + \frac{\Delta V}{R_3}, \quad \frac{\Delta P}{R_{tot}} = \frac{\Delta P}{R_1} + \frac{\Delta P}{R_2} + \frac{\Delta P}{R_3}$$

and

$$\frac{1}{R_{tot}} = \frac{1}{R_1} + \frac{1}{R_2} + \frac{1}{R_3}, \quad \frac{1}{R_{tot}} = \frac{1}{R_1} + \frac{1}{R_2} + \frac{1}{R_3}$$

Conductance, g, is the reciprocal of resistance. Therefore the total conductance of a parallel network of resistors is simply the sum of each individual conductance.

$$g_{tot} = g_1 + g_2 + g_3, \quad g_{tot} = g_1 + g_2 + g_3$$

Capacitance

As we have seen, the potential difference between two points (ΔV) is the difference in energy per unit mass or charge between those two points and arises because of a difference in field strength. In the case of electrical potential difference, this difference in field strength is a result of a difference in charge (ΔQ) between those two points, such as across a cell membrane.

$$\Delta Q \propto \Delta V$$

or

$$\Delta Q = C\Delta V,$$

where C = capacitance ($Q \cdot V^{-1}$, F).

Consider a simple circuit containing a capacitor (Fig. 101). We begin with the capacitor completely uncharged. When the switch is closed the capacitor begins to charge. Initially, electrons flow from the cell and reach side *a*. Since they cannot cross the *dielectric*, they begin to accumulate. This results in the formation of an electric field, which repels electrons from plate *b*. These then move around the circuit.

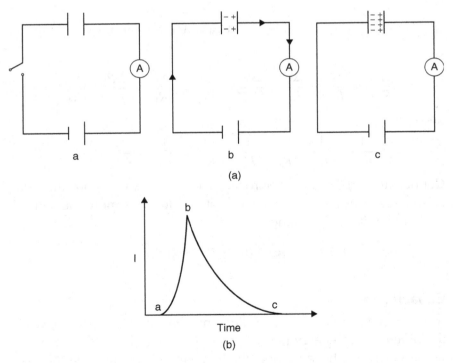

Fig. 101. Charging a capacitor.

Current is flow of charge, so a current is flowing and is recorded by the ammeter. Eventually, the strength of the electric field between the plates reaches a maximum (an equilibrium) and no more electrons are repelled from plate *b*. No more charge moves and so the current recorded by the ammeter drops to zero: the capacitative spike current has ceased and the capacitor is charged. Let us now consider what happens when the capacitor is allowed to discharge across the resistor.

If the First Law of Thermodynamic is to be obeyed, i.e. energy is not created or destroyed, the potential difference across the capacitor plates must at all times be equal and opposite to the potential difference across the resistor. Therefore:

$$V_{cap,t} = -V_{res,t},$$

where $V_{cap,t}$ = potential difference across capacitor at time *t* (V)
$V_{res,t}$ = potential difference across resistor at time *t* (V).

The potential difference across the capacitor plates is given by $\frac{Q}{C}$ and that across the resistor is given by *IR*, so:

$$\frac{Q_t}{C} = -I_t R$$

where Q_t = charge on capacitor at time *t* (C, A·s)
I_t = current through resistor at time *t* (C·s^{-1}, A).

Rearranging:

$$I_t = -\frac{Q_t}{RC}$$

Let $\tau = RC$, where τ = time constant (s). Then:

$$I_t = -\frac{Q_t}{\tau}$$

Since $I = \frac{dQ}{dt}$,

$$\frac{dQ_t}{dt} = -\frac{Q_t}{\tau}$$

and this integrates to give

$$Q_t = Q_{max}e^{-\frac{t}{\tau}}$$

where Q_{max} = maximum charge on capacitor (C, A/s).

Finally, since $Q = CV$ and C is constant we can write:

$$V_t = V_{max}e^{-\frac{t}{\tau}}$$

where V_{max} = maximum potential difference across capacitor (V). τ gives the time taken for the potential difference between the capacitor plates to fall by a factor of $\frac{1}{e}$. Clearly, altering either R or C will alter τ (Fig. 102).

It is instructive to think of a simple model of a capacitor (Fig. 103) and factors that influence C. For a parallel plate capacitor consisting of two plates of area A separated by a distance d:

$$C \propto \frac{A}{d}$$

where A = area of each plate (m^2)
      d = distance between plates (m).

The larger the area of the plates, the more space is available to store charge and so the more charge stored at a particular potential difference. The larger the distance between the plates, the

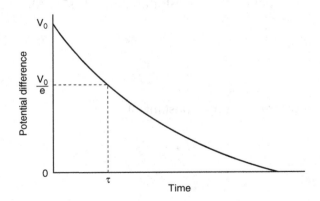

Fig. 102. Discharging a capacitor (decay of potential difference with time).

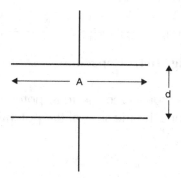

Fig. 103. The parallel plate capacitor.

weaker the electric field between the plates and the less charge stored at a particular potential difference. Introducing a constant of proportionality, ε:

$$C = \frac{\varepsilon A}{d}$$

where ε = permittivity (F·m^{-1}).

Permittivity is a property of the *dielectric*, and reflects the ability of the dielectric to shield against an electric field (it should not be confused with resistance: a material with a high permittivity does not necessarily have a high resistivity, i.e. is not necessarily a good insulator). The better the dielectric is at shielding against an electric field, the larger the field and therefore the amount of charge which can be stored on the plates before the capacitor spontaneously discharges (sparks). Incidentally, while it is possible to shield against electric fields it is not possible to shield against gravitational fields: this represents a fundamental difference (another is that positive and negative charge exist, while only positive mass exists).

We can now think about the total capacitance of a network of capacitors.

For capacitors arranged in series shown in Fig. 104, the same current must pass through each capacitor, so the charge stored on each capacitor must be the same:

$$V_1 = \frac{Q}{C_1}, \quad V_2 = \frac{Q}{C_2}, \quad V_3 = \frac{Q}{C_3}$$

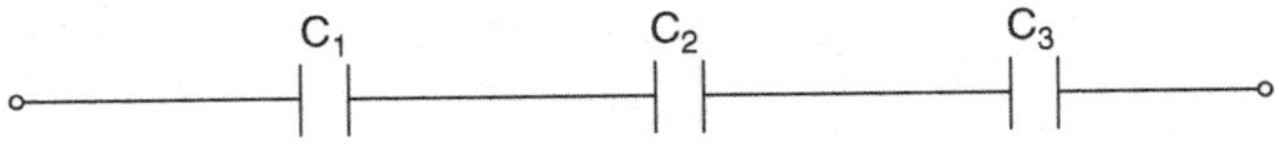

Fig. 104. Capacitors in series.

and if energy is to be conserved the total potential difference across all three capacitors must be the sum of the potential difference across each capacitor:

$$V_{tot} = V_1 + V_2 + V_3$$

so

$$\frac{Q}{C_{tot}} = \frac{Q}{C_1} + \frac{Q}{C_2} + \frac{Q}{C_3}$$

and

$$\frac{1}{C_{tot}} = \frac{1}{C_1} + \frac{1}{C_2} + \frac{1}{C_3}$$

Measuring Pressure

An understanding of the concept of pressure and the ways in which it can be measured is essential in all areas of physiology. The force exerted by a column of fluid on the bottom of its container (see Fig. 105) is given by:

$$F = mg_g$$

Expressing mass as the product of density and volume we then have:

$$F = \rho V g_g$$

Pressure is defined as the force acting per unit area. Therefore the pressure exerted by a column of fluid on the bottom of its container is independent of the cross-sectional area of the container. Here:

$$P = \rho g_g h$$

where h is the height of the column (m).

It follows that, if g_g is assumed to be constant, any pressure can be expressed as the height of a column of a fluid of a particular

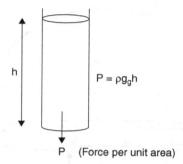

P (Force per unit area)

Fig. 105. Pressure exerted by a column of fluid.

density. This is the traditional means of expressing pressures in physiology and medicine. Mercury is a very dense material ($\rho_{mercury} = 13.6 \cdot 10^3$ kg·m^{-3}) and hence mmHg is a useful unit to express large pressures, such as arterial blood pressure. Smaller pressures, such as intra-alveolar pressure, are often measured in cmH$_2$O ($\rho_{water} = 10^3$ kg·m^{-3}). Therefore:

$$1\,\text{mmHg} = 1.36\,\text{cmH}_2\text{O} \quad \text{and} \quad 1\,\text{cmH}_2\text{O} = 0.76\,\text{mmHg}$$

Standard atmospheric pressure (realistic on a pleasant day at sea level) is 760 mmHg.

LaPlace's Law

LaPlace's law gives the relationship between the transmural tension experienced by a thin, curved elastic surface and its radius. It is effectively a statement of Newton's Third Law, "for every action there is an equal and opposite reaction". In all cases two forces act, an outward force due to a pressure difference across the wall and an inward force due to the transmural tension in the wall, expressed per unit length. We will consider the application of LaPlace's law to spheres (used as a model for alveoli and for the heart) and cylinders (used as a model for blood vessels).

Considering a sphere, these forces act as shown in Fig. 106.

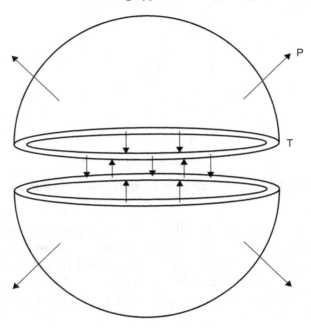

Fig. 106. Forces acting on a sphere such as an alveolus.

The outward force due to the transmural pressure difference acts perpendicular to the surface at all points on the inner surface.

$$F_{out} = (P_i - P_o) \cdot A,$$

where F_{out} = force acting outwards (N)
      P_i = pressure inside (Pa **or** mmHg)
      P_o = pressure outside (Pa **or** mmHg)
      A = area over which force acts, i.e. cross-sectional area (m^2).

Since for a sphere:

$$A = \pi r^2$$

it follows that:

$$F_{out} = (P_i - P_o)\pi r^2$$

The inward force can be thought of as the average tensile force acting along the circumference tending to pull the sphere together:

$$F_{in} = T \cdot C$$

where F_{in} = force acting inwards (N)
$\quad T$ = tension per unit length (N·m^{-1})
$\quad C$ = circumference (m).

Since for a sphere:

$$C = 2\pi r$$

it follows that:

$$F_{in} = T \cdot 2\pi r$$

Applying Newton's Third Law:

$$F_{in} = F_{out}$$

so

$$T \cdot 2\pi r = (P_i - P_o)\pi r^2$$

which rearranges to:

$$P_i - P_o = \frac{2T}{r}$$

By similar reasoning, the equation relating transmural pressure difference to transmural tension and radius for a cylinder can be shown to be:

$$P_i - P_o = \frac{T}{r}$$

Cell Volume and the Gibbs-Donnan Equilibrium

In Chapter 1, we mention the importance of extracellular Na^+ and of the Na^+/K^+-ATPase in the maintenance of cell volume. It is interesting to explore this further.

Let us consider a simple cell with a solution K^+, Cl^- and $A^{-1.2}$ in water inside and a solution of K^+ and Cl^- in water outside (Fig. 107). The cell membrane is permeable to K^+, Cl^- and water only.

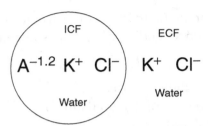

Fig. 107. Model cell.

The concentrations of K^+ and Cl^- will equilibrate across the cell membrane generating Nernst potential differences. Naturally $A^{-1.2}$ cannot contribute.

$$E_K = \frac{RT}{zF} \ln \left(\frac{[K^+]_o}{[K^+]_i} \right)$$

$$E_{Cl} = \frac{RT}{zF} \ln \left(\frac{[Cl^-]_i}{[Cl^-]_o} \right).$$

At equilibrium, the system will reach the lowest possible energy state such that:

$$E_K = E_{Cl}$$

so

$$\frac{RT}{zF} \ln \left(\frac{[K^+]_o}{[K^+]_i} \right) = \frac{RT}{zF} \ln \left(\frac{[Cl^-]_i}{[Cl^-]_o} \right)$$

$$\frac{[K^+]_o}{[K^+]_i} = \frac{[Cl^-]_i}{[Cl^-]_o}$$

$[K^+]_o[Cl^-]_o = [K^+]_i[Cl^-]_i$ (Gibbs-Donnan condition)

For simplicity later, let us define:

$$a = [K^+]_o, \quad b = [Cl^-]_o, \quad c = [K^+]_i, \quad d = [Cl^-]_i$$

So we have:

$$ab = cd$$

When a stable cell volume is reached, there is no net flow of water across the cell membrane. This must mean that the intracellular osmolarity is the same as the extracellular osmolarity, i.e.

$$[K^+]_o + [Cl^-]_o = [K^+]_i + [Cl^-]_i + [A^{-1.2}]_i$$

and since $[A^{-1.2}]_i$ must be positive

$$[K^+]_o + [Cl^-]_o < [K^+]_i + [Cl^-]_i$$

i.e.

$$a + b < c + d \text{ (osmotic condition)}.$$

The bulk extracellular solution must also have no net charge so

$$[K^+]_o = [Cl^-]_o \text{ (electroneutrality condition)}$$

i.e. $a = b$. Taking the osmotic and electroneutrality conditions together, we have:

$$2a < c + \delta$$

Squaring both sides:

$$(2a)^2 < (c + \delta)^2$$

$$4a^2 < c^2 + \delta^2 + 2c\delta$$

Both c and d must be positive so:

$$c^2 + \delta^2 + 2c\delta > 0$$

and thus

$$c^2 + \delta^2 > 2c\delta$$

so

$$c^2 + \delta^2 + 2c\delta > 4c\delta.$$

Combining this with $4a^2 < c^2 + \delta^2 + 2c\delta$:

$$4a^2 < 4c\delta$$

so

$$a^2 < c\delta$$

or $ab < c\delta$ i.e.

$$[K^+]_o[Cl^-]_o < [K^+]_i[Cl^-]_i$$

This disagrees with the Gibbs-Donnan condition which means that a steady state cannot be reached with a positive value of $[A^{-1.2}]_i$. The implication is that water must continue to enter the cell until $[A^{-1.2}]_i = 0$, i.e. the cell volume is infinite. Cells should expand until they burst. Happily, this does not happen in real life! This is because there is also a membrane-impermeant ion in the extracellular space. This exerts an osmotic effect equal and opposite to that of intracellular $A^{-1.2}$. It is worthwhile noting that the charge on this substance is irrelevant. In living organisms this is achieved by making the cell membrane impermeable to Na^+, an abundant ion in the extracellular fluid.

Of course, a biological membrane can never achieve absolute impermeability to any substance. Given that there is a huge Na^+ concentration gradient across cell membranes, it is inevitable that there will be some leak into the cell. Indeed, Na^+ is intentionally allowed to cross the cell membrane as part of a number of vital transport processes (see Chapter 2). The most important role of the Na^+/K^+-ATPase is to counteract this leak. If the Na^+/K^+-ATPase is poisoned, a cell will indeed swell and eventually burst.

Simplifying the Cable Equation

In Chapter 1 (Further Thoughts), we derive the cable equation:

$$V = \frac{r_m}{r_i}\frac{d^2 V}{dx^2} - r_m C_m \frac{dV}{dt}$$

To make this somewhat easier to work with we can define:

$$\lambda^2 = \frac{r_m}{r_i} \quad \text{or} \quad \lambda = \sqrt{\frac{r_m}{r_i}}$$

and

$$\tau = r_m C_m$$

so now we have:

$$V = \lambda^2 \frac{d^2 V}{dx^2} - \tau \frac{dV}{dt}$$

Making the assumption that an action potential is a constant current applied transversely across the cable for an infinitely long time, i.e. $\frac{dv}{dt} = 0$ we can neglect the capacitance term in the cable equation.

This then simplifies to:

$$V = \lambda^2 \frac{d^2 V}{dx^2}$$

Integrating twice with respect to t we have:

$$V_x = V_o e^{\frac{-x}{\lambda}}$$

and this is a simple exponential decay function (V successively halves over a constant distance, Fig. 108).

When $x = \lambda$ we have:

$$V_\lambda = V_o e^{\frac{-\lambda}{\lambda}}$$
$$V_\lambda = \frac{V_o}{e}$$

So we have ended up defining the space constant, λ, as the distance over which V falls to $\frac{1}{e}$ times its initial value. It can be seen that the distance of which V falls to half its initial value is given by $\ln 2 \cdot \lambda$. If the current that charged the membrane is suddenly switched off, V will decrease as c_m discharges. The time course over which this

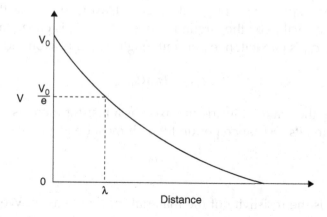

Fig. 108. Decay of potential difference with distance.

occurs is again a simple exponential function:

$$V_t = V_o e^{\frac{-x}{\tau}}$$

where τ is the time constant, i.e is the time over which V falls to $\frac{1}{e}$ times its initial value. It can again be seen that the time over which V falls to half its initial value is given by $\ln 2 \cdot \tau$.

Since the conductance (the reciprocal of resistance, measured in siemens, S) of a membrane must be proportional to its surface area, it is often convenient to express conductance per unit area, i.e. in $S \cdot m^{-2}$. Resistance is the reciprocal of conductance and is therefore inversely proportional to surface area, so somewhat counter-intuitively membrane resistance can be expressed in $\Omega \cdot m^2$. If we want to know the membrane resistance per unit length of axon, we can simply divide this value by the surface area per unit length, i.e. the circumference if the axon is cylindrical. Here R_m is resistance per unit area, r_m is resistance per unit length and a is the axon radius.

$$r_m = \frac{R_m}{2\pi a}$$

For membrane capacitance, the situation is different. Since capacitance is directly proportional to surface area, it can then be directly expressed per unit area, i.e. in $F \cdot m^{-2}$. If we want to know the membrane capacitance per unit length of axon, we must this time multiply this value by the circumference. Here C_m is capacitance per unit area, c_m is capacitance per unit length and a is axon radius.

$$c_m = 2\pi a C_m$$

If we treat the material inside the axon as a resistor then, as we know from earlier, its resistance per unit length must be given by:

$$r_i = \frac{\rho_i}{\pi a^2}$$

where ρ_i is the resistivity of the material inside the axon. We already know that $\lambda = \sqrt{\frac{r_m}{r_i}}$ so we can now substitute these values into this expression. We have:

$$\lambda = \sqrt{\frac{\frac{R_m}{2\pi a}}{\frac{\rho_i}{\pi a^2}}} = \sqrt{a}\sqrt{\frac{2R_m}{\rho_i}}$$

Importantly, this means that if R_m and ρ_i are constant then the space constant is directly proportional to the square root of the axon radius.

Similarly, for capacitance we already know that $\tau = r_m c_m$ and we can also substitute the expressions we have derived into this expression. This time we have:

$$\tau = \frac{R_m}{2\pi a} \cdot 2\pi a C_m = R_m C_m,$$

meaning that the time constant is not affected by axon radius.

Returning to the expression:

$$I_{trans} = \frac{1}{r_i} \cdot \frac{d^2 V}{dx^2}$$

and substituting in

$$r_i = \frac{\rho_i}{\pi a^2}$$

we have

$$i_{trans} = \frac{\pi a^2}{\rho_i} \cdot \frac{d^2 V}{dx^2}$$

Here, i_{trans} is a current per unit length and therefore I_{trans}, the current per unit area, must be given by:

$$I_{trans} = \frac{i_{trans}}{2\pi a}$$

and putting this into the previous expression, we have

$$I_{trans} = \frac{a}{2\rho_i} \frac{d^2 V}{dx^2}$$

As:

$$\frac{dV}{dx} = \frac{dt}{dx} \cdot \frac{dV}{dt} \quad \text{(chain rule)}$$

and $\frac{dx}{dt}$ is velocity, if θ is the conduction velocity then

$$\frac{dV}{dx} = \frac{1}{\theta} \cdot \frac{dV}{dt}$$

and

$$\frac{d^2V}{dx^2} = \frac{1}{\theta^2} \cdot \frac{d^2V}{dt^2}$$

Putting this into our equation for I_{trans}, we have

$$I_{trans} = \frac{a}{2\rho_i\theta^2} \cdot \frac{d^2V}{dt^2}$$

Assuming that I_{trans}, ρ_i and $\frac{d^2V}{dt^2}$ remain constant then:

$$\theta^2 \propto a$$

and

$$\theta \propto \sqrt{a}$$

It must be remembered, however, that since the assumptions required in this derivation are not really justified, this argument does not apply for *real* action potentials.

Index